APPROXIMATION BY POLYNOMIALS WITH INTEGRAL COEFFICIENTS

MATHEMATICAL SURVEYS · *Number 17*

APPROXIMATION BY POLYNOMIALS WITH INTEGRAL COEFFICIENTS

BY

LE BARON O. FERGUSON

1980

AMERICAN MATHEMATICAL SOCIETY

PROVIDENCE, RHODE ISLAND

Library of Congress Cataloging in Publication Data

Ferguson, Le Baron O 1939–

Approximation by polynomials with integral coefficients.

(Mathematical surveys; no. 17)

1. Approximation theory. 2. Polynomials. I. Title. II. Series: American Mathematical Society. Mathematical surveys; no. 17.

QA221.F46 511′.4 79-20331

ISBN 0-8218-1517-2

1980 Mathematics Subject Classifications. Primary 41A10, 41A29, 41A30, 41A25.

Printed in the United States of America

To the memory of SEABURY COOK

TABLE OF CONTENTS

PREFACE

Results in the approximation of functions by polynomials with coefficients which are integers have been appearing since that of Pál in 1914. The body of results has grown to an extent which seems to justify the present book. The intention here is to make these results as accessible as possible.

Aside from the intrinsic interest to the pure mathematician, there is the likelihood of important applications to other areas of mathematics; for example, in the simulation of transcendental functions on computers. In most computers, fixed point arithmetic is faster than floating point arithmetic and it may be possible to take advantage of this fact in the evaluation of integral polynomials to create more efficient simulations. Another promising area for applications of this research is in the design of digital filters. A central step in the design procedure is the approximation of a desired system function by a polynomial or rational function. Since only finitely many binary digits of accuracy actually can be realized for the coefficients of these functions in any real filter the problem amounts (to within a scale factor) to approximation by polynomials or rational functions with integral coefficients. For more details one may consult this author's listing in the Bibliography. It would be gratifying to the author if this book stimulates research in this direction.

Most of the results here have already appeared in the literature. However, for the expert, we mention the following exceptions: Corollaries 7.17, 7.20, Propositions 7.16, 9.8, and Theorems 9.7, 9.9, 9.10, 9.11, A.4, A.5.

It is a pleasure to acknowledge the help of many people in the writing of this book. It was my advisor, Edwin Hewitt, who initially brought the problem to my attention. G. G. Lorentz suggested the book itself. In learning the subject, especially as it relates to number theory, I am indebted to a number of valuable conversations with David Cantor. I would also like to express my gratitude for the support of the institutions listed at the end of the Bibliography and to the Air Force Office of Scientific Research for partial support from grants numbered AFOSR 71-2030 and AFOSR 78-3599. Finally, I thank Mrs. Joyce Kepler for her excellent services as typist.

Riverside
February, 1976

La vie est brève:
Un peu d'espoir,
Un peu de rêve
Et puis–bonsoir! Leon Montenaeken

INTRODUCTION

As an introduction to our subject we consider some elementary results and their simple proofs. Besides giving an indication of the kind of results to expect, they may be useful in themselves. Also, the techniques of proof will occur again in establishing the stronger results.

For the present an integral polynomial is a polynomial whose coefficients all lie in the set of rational integers $\{0, \pm 1, \pm 2, \ldots\}$. In references to the bibliography, we give the author's name, followed by the last two digits of the year of publication in square brackets. For references which appeared in the nineteenth century, all four digits are given.

The results in the theory of approximation by integral polynomials can be summarized very roughly as follows. In contrast with the classical case of arbitrary coefficients for the polynomials, approximation on a set X by integral polynomials is only possible if certain conditions are satisfied by the function to be approximated and the set X. The set X must not be too large in the sense that its transfinite diameter must be less than unity. If S has transfinite diameter less than unity, then there is a finite subset $J(X)$ of X such that uniform approximation to a continuous f is possible by integral polynomials if and only if f can be interpolated on $J(X)$ by such polynomials. Apparently the first result concerning the approximation of functions by integral polynomials is the following by Pál **[14]**. Let f be a continuous real valued function on an interval $[-\alpha, \alpha]$ with $0 < \alpha < 1$. Then f can be uniformly approximated by integral polynomials if and only if $f(0)$ is an integer. This is easily proved as follows (Ferguson **[70b]**). The condition that $f(0)$ be an integer is obviously necessary. Indeed, if k is an integer and $\{g_n\}$ a sequence of polynomials with integral coefficients tending uniformly to f on a set containing k, then $g_n(k) \to f(k)$ as $n \to \infty$. But each $g_n(k)$ is an integer; hence $f(k)$ is a limit point of the set of integers, hence an integer itself. Conversely, suppose $f(0)$ is an integer. Since it suffices to approximate $f - f(0)$ we can assume $f(0) = 0$. Let $\varepsilon > 0$. Since $0 < \alpha < 1$, $\Sigma_{n=1}^{\infty} \alpha^n < \infty$ and there is an odd integer k such that

$$\sum_{n>k} \alpha^n < \varepsilon/3. \tag{1}$$

Since k is odd, the function x^k separates the points of $[-\alpha, \alpha]$ and by the Stone-Weierstrass theorem there is a polynomial p_0 with real coefficients such that if $p(x) = p_0(x^k)$, $-\alpha \leqslant x \leqslant \alpha$, then

$$\|f - p\| < \varepsilon/3 \tag{2}$$

where $\|\cdot\|$ is the norm defined by $\|h\| = \sup_{|x|\leqslant\alpha}|h(x)|$. If we let p_1 be the polynomial p without its constant term, we see from the assumption $f(0) = 0$ and (2) that

$$\|p - p_1\| < \varepsilon/3. \tag{3}$$

Finally, if we define $[p_1]$ to be the polynomial p_1, with each coefficient replaced by its integral part, then $p_1 - [p_1]$ is a polynomial without constant term which involves only powers $\geqslant k$ and with coefficients between 0 and 1; hence by (1)

$$\|p_1 - [p_1]\| < \varepsilon/3. \tag{4}$$

From (2), (3), and (4) and the triangle inequality we have

$$\|f - [p_1]\| < \varepsilon$$

which establishes Pál's result.

It is natural to ask next what happens in case $\alpha = 1$. As we have noted, a continuous function which is approximable in the above sense must take on integral values at -1, 0, and 1. This is not a sufficient condition, however. Indeed, later that same year Kakeya **[14]** published the following generalization of Pál's result: a continuous real valued function f on $[-1, 1]$ is uniformly approximable by integral polynomials if and only if $f(-1)$, $f(0)$, and $f(1)$ are integers and $f(-1) + f(1)$ is even. The necessity of the latter condition is easily seen when one notes that if p is an integral polynomial, then $p(-1) + p(1)$ is twice the sum of the coefficients of the even powered monomials in p. As α tends upward to 2 we will see that, in order to be approximable, a continuous function needs to satisfy more and more conditions of an arithmetic nature. The number of conditions tends to infinity as α tends to 2.

When the polynomials are allowed to have arbitrary real coefficients then we know from Weierstrass' theorem that any continuous function can be uniformly approximated on any closed bounded interval. In the case of approximation by integral polynomials there are two major differences. First, as we have seen, only those functions which satisfy certain arithmetic conditions are approximable. The second difference is that in approximation by integral polynomials the set on which the approximation is to take place may be so "large" that the problem is trivial. Indeed, Kakeya **[14]** showed that on any interval of length $\geqslant 4$ no function can be uniformly approximated by integral polynomials unless it is identically equal to such a polynomial. This is easily proved as follows.

Suppose $\{p_n\}$ is a sequence of integral polynomials tending uniformly to a function f on $[a, b]$ which is not identically equal to an integral polynomial there. Then there exist n and m such that $\|p_n - f\| < \frac{1}{2}$, $\|p_m - f\| < \frac{1}{2}$ ($\|\cdot\|$ is the uniform norm on $[-1, 1]$), and $p_n \not\equiv p_m$. It follows that $\|p_n - p_m\| < 1$ and since $p_n - p_m$ is not zero it has a leading coefficient c, say, which is a nonzero integer; hence $|c| \geqslant 1$ and

$$\|(p_n - p_m)/c\| < 1. \tag{5}$$

However, $(p_n - p_m)/c$ is a monic polynomial (i.e., has leading coefficient unity); hence (5) is impossible. Indeed, it is a well-known result of Čebyšev (Lorentz [**66**, Chapter 2, Theorem 11]) that the monic polynomial of degree n ($n \geqslant 1$) which has least supremum norm on $[-1, 1]$ has the form $2^{1-n} \cos(n \times \cos^{-1} x)$, $-1 \leqslant x \leqslant 1$. It follows that the polynomial with the same attributes on $[-2, 2]$ has the form $2 \cos(n \cos^{-1}(x/2))$, $-2 \leqslant x \leqslant 2$. For any positive integer n these polynomials all have norm 2. Since translation does not change the norm or the monicity of a polynomial, it follows that every nonconstant monic polynomial on an interval of length greater than or equal to 4 has norm at least 2.

In 1925 the following result by Chlodovsky [**25**] appeared: if $[a, b]$ is an interval not containing an integer, then any continuous function f on $[a, b]$ can be uniformly approximated by integral polynomials. This is an immediate consequence of Pál's result but the proof is different. We first note that after translating by an integer we can make $0 < a < b < 1$; hence we assume this without loss of generality. Next notice that from Weierstrass' theorem it suffices to show that any constant can be uniformly approximated on $[a, b]$ by integral polynomials. (First approximate f by a polynomial with real coefficients and then replace each coefficient by an approximating integral polynomial.) Since every real number can be approximated by one of the form $n2^{-m}$ where n and m are integers, it suffices to approximate the constant $\frac{1}{2}$. But for large k the constant $\frac{1}{2}$ is uniformly approximated on $[a, b]$ by a function of the form $1/(2 - x^k)$. Finally

$$\frac{1}{2 - x^k} = \frac{1}{1 - (x^k - 1)} = \sum_{n=0}^{\infty} (x^k - 1)^n$$

where the series converges uniformly on $[a, b]$. Since any truncation of the series is a polynomial with integral coefficients we have Chlodovsky's result.

A final indication of some of the techniques of proof in this subject is the following which appeared in a paper by Kantorovič [**31**]: a continuous real valued function f on $[0, 1]$ is uniformly approximable by integral polynomials if and only if $f(0)$ and $f(1)$ are integers. We have already seen the necessity of this condition. Conversely, suppose the condition holds. Since it suffices to approximate $f(x) - (f(1)x + f(0)(1 - x))$, we can assume that $f(0) = f(1) = 0$. It is well known that the sequence of Bernšteĭn polynomials for f converges uniformly to f on $[0, 1]$ (Lorentz [**66**, Chapter 1, Theorem 4]) hence it suffices to approximate

$$p_n(x) = \sum_{\nu=1}^{n-1} f\left(\frac{\nu}{n}\right)\binom{n}{\nu} x^{\nu}(1 - x)^{n-\nu}$$

for all sufficiently large n. The $\nu = 0$ and $\nu = n$ terms are not present here since we have assumed that $f(0) = f(1) = 0$. Let

$$q_n(x) = \sum_{\nu=1}^{n-1} \left[f\left(\frac{\nu}{n}\right)\binom{n}{\nu} \right] x^\nu (1-x)^{n-\nu},$$

where $[\cdot]$ represents the greatest integer function, i.e., $[x]$ is the greatest integer $\leqslant x$. Since $\binom{n}{\nu} \geqslant n$ $(1 \leqslant \nu \leqslant n-1)$ we have

$$\sum_{\nu=1}^{n-1} x^\nu (1-x)^{n-\nu} \leqslant \frac{1}{n} \sum_{\nu=1}^{n-1} \binom{n}{\nu} x^\nu (1-x)^{n-\nu}$$
$$\leqslant \frac{1}{n} \sum_{\nu=0}^{n} \binom{n}{\nu} x^\nu (1-x)^{n-\nu} = \frac{1}{n},$$

by the binomial theorem. Thus, with $\|\cdot\|$ denoting the uniform norm on $[0, 1]$, we have $\|p_n - q_n\| \leqslant 1/n$ and since q_n is an integral polynomial, we are done.

In what follows we will establish generalizations of the above results as well as some related results. In order to be able to describe them economically we will first introduce some notation. The results fall into two main categories. On the one hand they characterize those functions that can be approximated uniformly or in the L_p norms by integral polynomials. These we call qualitative results. Some examples are those results already mentioned. On the other hand there are the quantitative results which give estimates of the rates of convergence of integral polynomials of best approximation. An example is the result of Kantorovič which appears later in the introduction.

Throughout, X will denote a compact Hausdorff space and for any subset $S \subset X$ we will use $\|\cdot\|_S$ to denote the uniform (Čebyšev, sup) norm on S. Thus for any bounded, real or complex valued function f on S, we have

$$\|f\|_S = \sup_{x \in S} |f(x)|.$$

The interior of X will be denoted $X°$.

The algebra of all complex continuous functions in this norm is denoted by $C(X)$, and the subalgebra of real valued elements by $C(X, R)$. We often write $\|\cdot\|$ in place of $\|\cdot\|_X$. The symbols **N**, **Z**, **Q**, **R**, and **C** are used to denote, respectively, the natural numbers $\{0, 1, 2, \ldots\}$, the rational integers $\{0, \pm 1, \pm 2, \ldots\}$, the rational numbers, the real numbers, and the complex numbers. A *monic polynomial* is one whose leading coefficient is unity. If A and B are two sets, the relative complement of B in A is denoted by $A \setminus B$. The empty set is denoted by $\varnothing$. By *integral polynomials* we mean polynomials with integral coefficients, where "integral" is the adjectival form of "integer." By *integers* we mean the elements of some discrete subring R of the complex numbers **C**. If $R = \mathbf{Z}$ we speak of *rational integers*. If $R = \mathbf{Z} + i\mathbf{Z}$ we have the so-called *Gaussian integers*. For a real number x, $[x]$ will denote the greatest integer which is $\leqslant x$, and (x) will denote the fractional part, $x - [x]$, of x.

If X is an interval of the real line **R** and f is a real or complex valued function defined on X, then we define

$$E_n(f) = \inf_{\deg p < n} \|f - p\|_X$$

where the polynomials p have real coefficients and

$$E_n^e(f) = \inf_{\deg q < n} \|f - q\|_X$$

where the polynomials q have rational integral coefficients. As a rule we reserve the symbol q for polynomials having integral coefficients.

As an example of a quantitative result we mention the following (Kantorovič [**31**]): if f is a continuous function on $X = [0, 1]$ with $f(0) = f(1) = 0$, then

$$E_n^e(f) \leqslant 2E_n(f) + O(1/n). \tag{6}$$

This can be proved as follows. Let n be any positive integer. Then there exists a (unique) polynomial p_n with degree $\leqslant n$ and real coefficients such that ($\|\cdot\| = \|\cdot\|_{[0,1]}$)

$$\|f - p_n\| = E_n(f). \tag{7}$$

Since $f(0) = 0 = f(1)$ we see from this that $|p_n(0)| \leqslant E_n(f) \geqslant |p_n(1)|$; hence

$$|p_n(1)x + p_n(0)(1 - x)| \leqslant E_n(f), \qquad 0 \leqslant x \leqslant 1.$$

Setting $\tilde{p}_n(x) = p_n(x) - (p_n(1)x + p_n(0)(1 - x))$ we have

$$\tilde{p}_n(0) = 0 = \tilde{p}_n(1) \tag{8}$$

and

$$\|p_n - \tilde{p}_n\| \leqslant E_n(f). \tag{9}$$

Thus by (7) and (9)

$$\|f - \tilde{p}_n\| \leqslant 2E_n(f). \tag{10}$$

It is easy to see that any polynomial of degree at most n can be written as a linear combination of the terms $\{x^\nu(1 - x)^{n-\nu}\}_{\nu=0}^n$; hence for some choice of real numbers a_ν $(0 \leqslant \nu \leqslant n)$ we have

$$\tilde{p}_n(x) = \sum_{\nu=0}^{n} a_\nu x^\nu (1 - x)^{n-\nu}$$

and by (8)

$$\tilde{p}_n(x) = \sum_{\nu=1}^{n-1} a_\nu x^\nu (1 - x)^{n-\nu}.$$

As above, if we set $[\tilde{p}_n](x) = \sum_{\nu=1}^{n-1}[a_\nu]x^\nu(1 - x)^{n-\nu}$ then we have $\|\tilde{p}_n - [\tilde{p}_n]\| \leqslant 1/n$. This together with (10) gives

$$\|f - [\tilde{p}_n]\| \leqslant 2E_n(f) + 1/n.$$

This establishes (6). We note in passing that much stronger results are known. See Chapter 12.

The qualitative results are divided into four cases, as follows. If X is a compact subset of $\mathbf{R}$ and $R = \mathbf{Z}$ we say that we are in the *real case*. If R is an arbitrary but fixed discrete subring of **C** with rank 2 and X a compact subset of $\mathbf{C}^n$ we say that we are in the *complex case*. The most complete results in this case hold for $n = 1$. A more general case is the following. For lack of a better word we call it the *adelic case*. Let T be a finite set of equivalence classes of valuations on an algebraic number field K which contains all the Archimedean classes. For each v in T let K_v be the corresponding completion of K, X_v a compact subset of K_v, and f_v a K_v-valued continuous function on X_v. The question is whether or not there exists p in $K[x]$ with T-integers for coefficients and such that $p - f_v$ is uniformly small on X_v for each v in T. The final case is that in which X is any compact Hausdorff space and $\mathfrak{F}$ is a point separating family of continuous functions on X. The integral polynomials in this case are $\mathbf{Z}[\mathfrak{F}]$, the polynomials in elements of $\mathfrak{F}$ with rational integral coefficients. We call this the *general case*. See Chapter 9.

We give criteria in all the above cases which characterize the functions which can be uniformly approximated by polynomials with integral coefficients. Proofs are given in all but the adelic case where, however, the results are stated completely and the connections with the previous cases are indicated. Although the adelic case could have been established first and the results of the real and complex cases derived from it, we have not done so because this would have limited the usefulness of the qualitative part of the book to those readers conversant with algebraic number theory.

PART I: PRELIMINARIES

CHAPTER 1

DISCRETE RINGS

In the real case we will take the rational integers for the coefficients of our integral polynomials. In the complex case, however, a larger ring is needed in order to include some nonreal numbers among the coefficients. It happens that we can establish our results for a whole class of rings: those which are discrete and have rank 2. We proceed to define these terms and to establish the properties of these rings which we will need later. We will also do the same for the adelic case.

DEFINITION 1.1. Throughout the following, A will denote a fixed but arbitrary discrete subring of $\mathbf{C}$ with rank 2. A subring of $\mathbf{C}$ which is discrete but not necessarily of rank 2 will usually be denoted by R. By *discrete* we mean that A is discrete as a subset of the topological space $\mathbf{C}$. By *rank* 2 we mean that the real linear space spanned by A has dimension 2.

A reader not interested in maximum generality may think of A as the ring of Gaussian integers $\mathbf{Z} + i\mathbf{Z}$.

The requirement that A have rank 2 is actually equivalent to A not being a subring of $\mathbf{R}$, as follows. Suppose A has rank less than 2 and $A \not\subseteq \mathbf{R}$. Then there exists $z \in A \setminus \mathbf{R}$. Thus z and z^2 are linearly dependent over $\mathbf{R}$; hence there exist $a, b \in \mathbf{R}$, not both zero, such that $az + bz^2 = 0$. Since $z \neq 0$ we have $a + bz = 0$. Since a and b are not both zero, this equation shows that neither is zero. Solving for z shows that $z \in \mathbf{R}$, a contradiction. Thus, if A has rank less than 2, then $A \subset \mathbf{R}$. The converse is obvious.

The requirement that a subring A of $\mathbf{C}$ be discrete is equivalent to the condition that $0 \neq z \in A$ implies $|z| \geqslant 1$. Indeed, if $0 < |z| < 1$ and $z \in A$, then $z^n \to 0$; hence 0 is a limit point of A and A is not discrete. Conversely the condition implies that $|z_1 - z_2| \geqslant 1$ for every distinct $z_1, z_2 \in A$. Hence the open unit disk centered at $z \in A$ meets A only in z which shows that z is an isolated point of A. Since z is any point of A, this shows that A is discrete.

The following result will be of use to us as we will often have occasion to replace the complex coefficients of an approximating polynomial by "nearest" elements of A.

PROPOSITION 1.2. *There is a* $\delta > 0$ *such that if* z *is any complex number, there exists* $a \in A$ *with* $|z - a| < \delta$.

PROOF. By Bourbaki [**63**, Theorem 1, p. 77], there exist b_1 and b_2 in **C** which are linearly independent over the reals and which generate A as an additive group. Since **C** has dimension 2 as a real vector space, there exist real numbers r_1 and r_2 such that $z = r_1 b_1 + r_2 b_2$. For $i = 1$ or 2 there exist integers n_i and real numbers r_i' such that $r_i = n_i + r_i'$ and $|r_i'| \leqslant \frac{1}{2}$. Then $n_1 b_1 + n_2 b_2$ is in A and $|z - (n_1 b_1 + n_2 b_2)| = |r_1' b_1 + r_2' b_2| \leqslant \frac{1}{2}(|b_1| + |b_2|)$. It suffices to set $\delta = \frac{1}{2}(|b_1| + |b_2|) + 1$. □

A *quadratic field* is a field F containing the rationals **Q** such that the degree of F over **Q**, denoted by $[F : \mathbf{Q}]$, is equal to two. It is well known (Weiss [**63**]) that every quadratic field F is of the form $\mathbf{Q}(\sqrt{d})$ for a unique square-free rational integer d different from 0 and 1. If d is negative (positive) the field $\mathbf{Q}(\sqrt{d})$ is said to be *imaginary* (*real*). Throughout this survey the symbol L will always denote an imaginary quadratic field. There is a connection between discrete rings of rank 2 and imaginary quadratic fields which we proceed to establish.

DEFINITION 1.3. If R is any subring with identity of the complex numbers **C**, then we say that an element z of **C** is *integral over* R if z is a root of a monic polynomial with coefficients in R.

PROPOSITION 1.4. *If* F *is a quadratic field, the elements of* F *which are integral over* **Z** *form a ring containing* **Z**.

PROOF. Jacobson [**51**, Theorem 8, p. 182]. □

DEFINITION 1.5. The ring of elements of a quadratic field F which are integral over **Z** is denoted by I_F. When we have some definite quadratic field F in mind, we often use the term *integer* to mean an element of I_F. Throughout the following, elements of the ring **Z** will be referred to as *rational integers*.

PROPOSITION 1.6. *Let* $F = \mathbf{Q}(\sqrt{d})$ *be a quadratic field. Then a basis for* I_F *as a* **Z**-*module is given by*

(i) $\{1, \sqrt{d}\}$ *if* $d \not\equiv 1 \pmod 4$, *and*

(ii) $\{1, (1 + \sqrt{d})/2\}$ *if* $d \equiv 1 \pmod 4$.

PROOF. By Jacobson [**51**, Theorem 2, p. 186], if $d \not\equiv 1 \pmod 4$, then $I_L = \{m + n\sqrt{d} : m, n \in \mathbf{Z}\}$ and it is easily seen that 1 and $\sqrt{d}$ are linearly independent over the rational integers **Z**. Likewise, if $d \equiv 1 \pmod 4$, the same theorem in Jacobson shows that

$$I_L = \{m + n\sqrt{d} : m, n \in \mathbf{Z}\} \cup \left\{\tfrac{1}{2}((2m + 1) + (2n + 1)\sqrt{d}) : m, n \in \mathbf{Z}\right\}.$$

Plainly this set is just $\{m + n(1 + \sqrt{d})/2 : m, n \in \mathbf{Z}\}$. To see that (ii) is actually linearly independent over **Z**, suppose that $m, n \in \mathbf{Z}$ and that

$$m + n(1 + \sqrt{d})/2 = 0. \qquad (*)$$

We must show that both m and n are zero. This is clear if $n = 0$. But if $n \neq 0$,

(*) shows that $\sqrt{d}$ is rational, which we know to be false since $[F : \mathbf{Q}] = 2$. □

PROPOSITION 1.7. *If L is an imaginary quadratic field, then the ring I_L is discrete and has rank* 2; *that is, I_L satisfies the conditions on A in Definition* 1.1.

PROOF. Write $L = \mathbf{Q}(\sqrt{d})$, as usual. Then $d < 0$, which implies that $\sqrt{d}$ is purely imaginary. This shows that the bases in Proposition 1.6 are linearly independent over $\mathbf{R}$, so I_L has rank 2. From this linear independence we also see that I_L is discrete (Bourbaki [**63**, pp. 74–75]). □

We note in passing that if F is a real quadratic field, then I_F does not satisfy the conditions on A in Definition 1.1 as follows. Since F is real, $F \subset \mathbf{R}$ and then $I_F \subset \mathbf{R}$, by definition. Thus F does not have rank 2 by the comment following Definition 1.1. Also, I_F is not discrete. Indeed, it is dense in $\mathbf{R}$ as follows. Let $F = \mathbf{Q}(\sqrt{d})$. Then a basis for I_F as a $\mathbf{Z}$-module is given by Proposition 1.6. The first element of each of these bases is unity and the second element is irrational; otherwise, $\sqrt{d}$ would be rational and $\mathbf{Q}(\sqrt{d}) = \mathbf{Q}$, a contradiction. Thus, by the well-known theorem of Kronecker, linear combinations with rational integer coefficients of these base elements are dense in $\mathbf{R}$.

PROPOSITION 1.8. *For quadratic fields $\mathbf{Q}(\sqrt{d_1})$ and $\mathbf{Q}(\sqrt{d_2})$, the nonequality $I_{\mathbf{Q}(\sqrt{d_1})} \neq I_{\mathbf{Q}(\sqrt{d_2})}$ implies $I_{\mathbf{Q}(\sqrt{d_1})} \cap I_{\mathbf{Q}(\sqrt{d_2})} = \mathbf{Z}$.*

PROOF. It is clear from Proposition 1.6 that $I_{\mathbf{Q}(\sqrt{d_1})} \cap I_{\mathbf{Q}(\sqrt{d_2})} \supset \mathbf{Z}$. Suppose that $z \in (I_{\mathbf{Q}(\sqrt{d_1})} \cap I_{\mathbf{Q}(\sqrt{d_2})}) \setminus \mathbf{Z}$. The proof now splits into cases. Suppose first that $d_1 \equiv d_2 \equiv 1 \pmod 4$. Then we have

$$z = m + n(1 + \sqrt{d_1})/2 = m' + n'(1 + \sqrt{d_2})/2$$

where m, n, m', n' are in $\mathbf{Z}$ and $n \neq 0 \neq n'$. Thus since $n \neq 0$,

$$r\sqrt{d_1} = r_1 + r_2\sqrt{d_2}, \qquad r_1, r_2 \in \mathbf{Q}.$$

This gives $\mathbf{Q}(\sqrt{d_1}) = \mathbf{Q}(r_1 + r_2\sqrt{d_2}) = \mathbf{Q}(r_2\sqrt{d_2}) = \mathbf{Q}(\sqrt{d_2})$ which implies $I_{\mathbf{Q}(\sqrt{d_1})} = I_{\mathbf{Q}(\sqrt{d_2})}$, a contradiction. The cases $d_1 \not\equiv 1 \equiv d_2 \pmod 4$ and $d_1 \equiv 1 \not\equiv d_2 \pmod 4$ (which are the same, by symmetry) and $d_1 \not\equiv 1 \not\equiv d_2 \pmod 4$ follow by the same argument. □

PROPOSITION 1.9. *If R is a discrete subring of $\mathbf{C}$, then $R \subset I_L$ for some imaginary quadratic field.*

PROOF [ADAPTED FROM PÓLYA [**23**, FOOTNOTE, p. 27]]. By Bourbaki [**63**, Theorem 1, p. 77] we have $R = \alpha\mathbf{Z}$ or $R = \alpha\mathbf{Z} + \beta\mathbf{Z}$ where $\alpha, \beta \in \mathbf{C}$ and in the second case α and β are linearly independent over $\mathbf{R}$. If $R = \alpha\mathbf{Z}$ we have $\alpha^2 = n\alpha$; so, $\alpha = n \in \mathbf{Z}$ unless $\alpha = 0$, in which case the conclusion is obvious. Then we have $R = n\mathbf{Z} \subset \mathbf{Z} \subset I_L$ for all L. If $R = \alpha Z + \beta Z$, we first show that $R \cap Z \neq \{0\}$. Since R is a ring we have

$$\alpha\beta = k\alpha + k'\beta, \tag{1}$$

$$\beta^2 = m\alpha + m'\beta, \tag{2}$$

$$\alpha\beta^2 = n\alpha + n'\beta \tag{3}$$

where $k, k', m, m', n, n' \in \mathbf{Z}$. If we multiply (1) by $m\alpha$ and (2) by $-k\alpha$ and add, we obtain $m\alpha^2\beta - k\alpha\beta^2 = (mk' - km')\alpha\beta$. Since α and β are linearly independent over $\mathbf{R}$, β is different from 0, and so

$$m\alpha^2 - k\alpha\beta = (mk' - m'k)\alpha. \tag{4}$$

If we multiply (1) by n and (3) by $-k$ and add, we find $n\alpha\beta - k\alpha\beta^2 = (nk' - kn')\beta$ or

$$n\alpha - k\alpha\beta = (nk' - kn'). \tag{5}$$

Combining (4) and (5) we have

$$m\alpha^2 + (km' - k'm - n)\alpha = kn' - k'n. \tag{6}$$

Thus $k'n - kn'$ is in R. If $k'n - kn' = 0$, then by (6), $km' - k'm - n$ is in R. If $km' - k'm - n$ is also zero, then by (6), $m = 0$, and then by (2), $0 \neq m' = \beta$, which is in $R \cap \mathbf{Z}$. Thus $R \cap \mathbf{Z} \neq \{0\}$. Since $R \cap \mathbf{Z}$ is an additive group, we can define $\rho = r\alpha + r'\beta$, $r, r' \in \mathbf{Z}$, to be the least positive member of $R \cap \mathbf{Z}$. Clearly $(r, r') = 1$ so there exists $s, s' \in Z$ such that

$$rs' - r's = \det\begin{vmatrix} r & r' \\ s & s' \end{vmatrix} = 1. \tag{7}$$

Thus the matrix in (7) transforms the basis $\{\alpha, \beta\}$ into a new basis $\{\rho, \sigma = s\alpha + s'\beta\}$ for R as a $\mathbf{Z}$-module, by Bourbaki **[63**, p. 75]. In particular, we have $\sigma^2 = t\rho + t'\sigma$, $t, t' \in \mathbf{Z}$, which shows that σ is integral over $\mathbf{Z}$ and has degree at most 2 over $\mathbf{Q}$. Then $\mathbf{Q}(\sigma)$ is a quadratic field, since the linear independence of $\{\alpha, \beta\}$ over $\mathbf{R}$ and (7) imply the linear independence of $\{\rho, \sigma\}$ over $\mathbf{R}$ and therefore $\mathbf{Q}$. Also, $\mathbf{Q}(\sigma)$ is imaginary since, if not, we would have $\mathbf{Q}(\sigma) \subset \mathbf{R}$ and σ and ρ would be linearly dependent over $\mathbf{R}$. It is clear that $R \subset I_{\mathbf{Q}(\sigma)}$. □

Our next result is the promised connection between the rings we are denoting by A and the imaginary quadratic fields.

PROPOSITION 1.10. *Let A be a discrete subring of* $\mathbf{C}$ *with rank* 2. *Then there is exactly one imaginary quadratic field L such that $A \subset I_L$.*

PROOF. The existence of L was shown in Proposition 1.9. Since A has rank 2, $A \not\subset \mathbf{Z}$ and uniqueness follows from Proposition 1.8. □

Not only is A contained in an I_L as above, but a kind of reverse inclusion also holds. Together these relationships will allow us to prove our theorems for I_L's in place of the more general A's.

PROPOSITION 1.11. *Let A be a discrete subring of rank* 2 *and $A \subset I_L$ as in Proposition* 1.10. *Then there exists a positive integer m_0 such that $m_0 I_L \subset A$.*

PROOF. Let A_1 be the smallest ring with identity containing A, i.e., $A_1 = \{n + a: n \in \mathbf{Z}, a \in A\}$. Let $0 \neq m \in A \cap \mathbf{Z}$ as in the proof of Proposition 1.9. Then $mA_1 \subset A$ since $m(n + a) = mn + ma \in A$. Now suppose $L = \mathbf{Q}(\sqrt{d})$ with $d \not\equiv 1 \pmod 4$. By Proposition 1.6, a basis for I_L as a $\mathbf{Z}$-module is $\{1, \sqrt{d}\}$.

Since A has rank 2 we can pick $z \in A \setminus \mathbf{Z}$. Since $A \subset I_L$, we can write $z = j + k\sqrt{d}$ where $j, k \in \mathbf{Z}$ and $k \neq 0$. Then $k\sqrt{d} \in A_1$. We claim that $mkI_L \subset mA_1$. Indeed, if $z \in I_L$, then $z = n + l\sqrt{d}$, $n, l \in \mathbf{Z}$, and $mkz = m(kn + lk\sqrt{d}) \in mA_1$ since $k\sqrt{d} \in A_1$. We have established $mkI_L \subset mA_1 \subset A$; hence it suffices to take $m_0 = mk$. The case $d \equiv 1 \pmod 4$ is handled similarly, taking $m_0 = 2mk$. □

PROPOSITION 1.12. *If $z_0 \in \mathbf{C}$ and z_0 is integral over I_L, then the minimal polynomial p of z_0 over L is in $I_L[z]$.*

PROOF. By hypothesis, z_0 satisfies a monic polynomial q in $I_L[z]$. Then p divides q and by Bourbaki [**64**, Proposition 11, p. 17], p is in $I_L[z]$. □

DEFINITION 1.13. Let F be any subfield of $\mathbf{C}$ with subring R with identity. Two algebraic numbers z_1 and z_2 are said to be *conjugate over F* if they have the same minimal polynomial over F. Given a monic irreducible polynomial $p \in F[z]$, the set of all its roots in $\mathbf{C}$ (say $\{z_1, \ldots, z_n\}$) is called a *complete set of conjugates over F*. In the case where p is in $R[z]$, we say that $\{z_1, \ldots, z_n\}$ is a *complete set of conjugates integral over R*.

In the adelic case our "integers" are defined as follows. Let K be any algebraic number field and Ω the set of equivalence classes of valuations on K. If $v \in \Omega$ we denote the canonical member of v by $|\cdot|_v$. That is, if the completion K_v of K under v is $\mathbf{R}$ or $\mathbf{C}$, then $|\cdot|_v$ is the usual or square of the usual absolute value on K_v, respectively, and if not, then K_v is a finite extension of a p-adic field $\mathbf{Q}_p$ and here $|\cdot|_v$ is normalized such that $|p|_v = p^{-N_v}$ where $N_v = [K_v : \mathbf{Q}_p]$.

DEFINITION 1.14. If T is any subset of Ω containing all the Archimedean elements of Ω, then we define the ring K^T of T-integers of K by

$$K^T = \{k \in K: |k|_v \leqslant 1 \text{ all } v \in \Omega \setminus T\}.$$

EXAMPLE 1.15. We note that if T consists of only the Archimedean elements of Ω then K^T consists of all elements of K which have norm $\leqslant 1$ for every non-Archimedean valuation on K. These are exactly the algebraic integers of K (O'Meara [**63**, §33 J]). For example, if $K = \mathbf{Q}$ then there is exactly one Archimedean valuation (up to equivalence) on $\mathbf{Q}$ (O'Meara [**63**, 12 : 1]) which we shall denote by ∞. Thus we can take $T = \{\infty\}$ and we will have $K^T = \mathbf{Z}$, the ordinary rational integers. Similarly, if $K = L$, an imaginary quadratic field, then $L = \mathbf{Q}(\sqrt{d})$ for $d < 0$ and by O'Meara [**63**, 15 : 8] L has just one Archimedean valuation (up to equivalence) which we again denote by ∞, and take T to be simply $\{\infty\}$. We then have that $K^T = I_L$.

CHAPTER 2

ČEBYŠEV POLYNOMIALS AND TRANSFINITE DIAMETER

The purpose of this chapter is to define and develop some basic properties of Čebyšev polynomials and transfinite diameter. We will see in Theorem 2.11 that if the transfinite diameter of a compact subset X of $\mathbf{C}$ is greater than or equal to unity then no function on X can be uniformly approximated by integral polynomials unless it is identically equal to such a polynomial on X.

We say that two polynomials are equal as functions if they give identically equal functions (written $p_1 \equiv p_2$) when restricted to the set X in question. We say they are equal as polynomials when they have the same coefficients.

DEFINITION 2.1. Let n be a nonnegative integer. We define $P_n = \{p \in \mathbf{C}[z]: p$ is monic and $\deg p = n\}$. By a *monic polynomial*, we mean a polynomial with a leading coefficient which is equal to 1.

PROPOSITION 2.2. *Let X be a compact subset of $\mathbf{C}$. For every nonnegative integer n there exists $t_n(z, X) \in P_n$ such that*

$$\|t_n(z, X)\|_X = \inf\{\|t\|_X: t \in P_n\}.$$

PROOF. Let $\alpha = \inf\{\|t\|: t \in P_n\}$. The problem of minimizing $\|z^n + a_{n-1}z^{n-1} + \cdots + a_0\|_X$, where the a's run through $\mathbf{C}$, is clearly equivalent to that of minimizing

$$\|z^n - (b_{n-1}z^{n-1} + \cdots + b_0)\|_X, \tag{$*$}$$

as the b's run through $\mathbf{C}$. We will show that there exist $b_0, \ldots, b_{n-1}$ making $(*)$ a minimum. Let M denote the subspace of $C(X)$ generated by $z^{n-1}, \ldots, z, 1$. We must find $p \in M$ such that $\|z^n - p(z)\| = \alpha$. We can restrict our attention to the case where $p \in S = \{g \in M: \|g\| \leq 2\|z^n\|\}$ since if $\|g\| > 2\|z^n\|$, then

$$\|z^n - g(z)\| \geq |\|z^n\| - \|g(z)\|| > \|z^n\|.$$

However, we have $\|z^n\| < \alpha$, unless $p = 0$ is already a solution of the problem. Since M is a finite dimensional topological vector space, it is locally compact. Thus S is compact. Also, the map $g \to \|z^n - g(z)\|$ is continuous, which implies that it attains its minimum over S. □

The sequence of polynomials $\{t_n\}$ is uniquely determined. This is a consequence of the fundamental result of Haar which follows in a generalized version. The set X can be any compact Hausdorff space here.

PROPOSITION 2.3. *If X has at least n points and $\varphi_1, \ldots, \varphi_n$ are elements of $C(X)$, then approximation by elements of their linear span in $C(X)$ is unique if and only if $a_1 = a_2 = \cdots = a_n = 0$ whenever $\Sigma_{j=1}^n a_j\varphi_j$ has n distinct zeros. In particular, as a function in $C(X)$ the polynomial $t_n(z, X)$ in Proposition 2.2 is uniquely determined.*

PROOF. We start with the first statement of the proposition. Let $V = \operatorname{span}\{\varphi_1, \ldots, \varphi_n\}$. Thus every $p \in V$ has the form $p = \Sigma_{j=1}^n a_j\varphi_j$. Suppose we have a $p \in V \setminus \{0\}$ with more than $n - 1$ distinct zeros. Let $f \in C(X)$ and set $m = \min_{p \in V}\|f - p\| = \|f - p^*\|$. The existence of p^* follows by essentially the same compactness argument as in the proof of Proposition 2.2. By the Hahn-Banach theorem we have

$$m = \max\{|\lambda(f)| : \lambda \in C^*(X), \|\lambda\| = 1, \lambda(\varphi_j) = 0\ (1 \leqslant j \leqslant n)\}$$

where $C^*(X)$ represents the dual of $C(X)$. From the w^*-compactness of the unit ball of $C^*(X)$ it follows that this maximum is attained, say for λ^*. We can assume that $\lambda^*(f) = m$. From the Riesz representation theorem (cf. Rudin **[74]**) there is a complex Borel measure μ^* with $\lambda^*(f) = \int f\, d\mu^*$, $f \in C(X)$, and $\int |d\mu^*| = 1$. Also, from the definition of λ^*, $\int p^*\, d\mu^* = 0, p \in V$. Thus

$$m = \int f\, d\mu^* = \int (f - p^*)\, d\mu^* \leqslant \|f - p^*\| \int |d\mu^*| = m;$$

hence

$$\int (f - p^*)\, d\mu^* = \left(\max_{x \in X} |f(x) - p^*(x)|\right) \int |d\mu^*|$$

which forces $f - p^* = \overline{(\operatorname{sgn} d\mu^*)} m$ on supp μ^*, the support of the measure μ^*.

Suppose the proposition is false. Then there exist distinct polynomials p_1 and p_2 of best approximation to f. As we have just seen $p_3 = p_2 - p_1 = 0$ at each point of supp μ^*. Since $p_3 \not\equiv 0$ there are at most $r \leqslant n - 1$ such points, say $x_1, \ldots, x_r$. Thus $\lambda^* = \Sigma_{j=1}^r c_j\delta_{x_j}$ where for any $x \in X$, δ_x represents the point evaluation functional ($\delta_x(f) = f(x)$, $f \in C(X)$) on $C(X)$ and the c_j's are constants. Since by hypothesis X has at least n points, we can choose $x_j \in X$, $r + 1 \leqslant j \leqslant n$, such that the points $x_1, \ldots, x_n$ are distinct and we have ($c_j = 0$, $r + 1 \leqslant j \leqslant n$) $\lambda^* = \Sigma_{j=1}^n c_j\delta_{x_j}$. Since $\lambda^* = 0$ on V the functionals $\delta_{x_1}, \ldots, \delta_{x_n}$ do not span all of the dual V^* of V. Also V is finite dimensional, hence reflexive. It follows that there is a polynomial $p \in V$, $p \not\equiv 0$, with $0 = \delta_{x_j}(p) = p(x_j)$, $1 \leqslant j \leqslant n$. Since the x_j are distinct, this contradicts our hypothesis; hence $p_1 = p_2$ and the approximation is unique.

To prove the other direction of the first statement of the proposition, we suppose that $p_1 \in V$ satisfies $p_1 \not\equiv 0$, $p_1(x_j) = 0$, $1 \leqslant j \leqslant n$, and the x_j are distinct and show that consequently the approximation fails to be unique. Since

$p_1 \not\equiv 0$, $\delta_{x_j}(p_1) = 0$, $1 \leqslant j \leqslant n$, and dim $V^* = n$, the δ_{x_j} are linearly dependent as members of V^*. Thus there is a linear combination $\lambda = \Sigma_{j=1}^n c_j \delta_{x_j}$ with not all $c_j = 0$ and $\lambda = 0$ on V. We can suppose that $\Sigma_{j=1}^n |c_j| = 1$. Thus $\|\lambda\| = \Sigma_{j=1}^n |c_j| = 1$ also, where $\|\lambda\|$ is the norm of λ as a linear functional on $C(X)$. Using Tietze's extension theorem it is easy to construct a function $g \in C(X)$ such that $\|g\| \leqslant 1$ and $g(x_j) = \overline{\text{sgn } c_j}$, $1 \leqslant j \leqslant n$. Choose $b > 0$ such that $\|bp_1\| < 1$ and set $f = g(1 - |bp_1|)$. We will show that f has (uncountably) many best approximations from V. For any $p \in V$ we have

$$1 = \sum_{j=1}^n |c_j| = |\lambda(f)| = |\lambda(f - p)| \leqslant \|\lambda\| \, \|f - p\| = \|f - p\|.$$

On the other hand for any $x \in X$ and $\varepsilon \in \mathbf{C}$ with $|\varepsilon| \leqslant 1$ we have

$$\begin{aligned} |f(x) - \varepsilon b p_1(x)| &\leqslant |f(x)| + |\varepsilon b p_1(x)| \\ &\leqslant 1 - |bp_1(x)| + |\varepsilon b p_1(x)| = 1 - (1 - |\varepsilon|)|bp_1(x)| \leqslant 1. \end{aligned}$$

Thus $\varepsilon b p_1$ is a best approximation to f from V for all ε with $|\varepsilon| \leqslant 1$ which shows that uniqueness of approximation fails. This completes the proof of the first statement.

The second statement of the proposition is a consequence of the first as follows. The determination of $t_n(z, X)$ amounts to approximating the function z^n by elements of the subspace of $C(X)$ spanned by the functions $1, z, \ldots, z^{n-1}$. But the latter are polynomials of degree at most $n - 1$, hence can have at most $n - 1$ distinct zeros without being identically zero. □

In view of the preceding two propositions we can make the following definition.

DEFINITION 2.4. If X is an infinite compact subset of $\mathbf{C}$, then the *nth Čebyšev polynomial for* X is the $t_n(z, X) \in P_n$ such that

$$\|t_n(z, X)\|_X = \inf\{\|t\|_X : t \in P_n\}.$$

If X contains exactly m elements, where m is finite, then we define $t_n(z, X)$ as above for $n \leqslant m$ and set $t_n(z, X) = \Pi_{x \in X}(z - x)$ for $n > m$. We frequently write $t_n(z, X)$ as $t_n(z)$ or t_n if there is no danger of confusion.

We note that $t_n(z, X)$ is well defined as a polynomial. Indeed, by 2.3, it is uniquely determined as a function. Thus if card X is infinite, then $t_n(z, X)$ is uniquely determined as a polynomial. On the other hand, if card $X = m < \infty$, then either $n \leqslant m$, in which case $t_n(z, X)$ is uniquely determined as a polynomial, or $n > m$, in which case the polynomial $t_n(z, X)$ is given explicitly.

PROPOSITION 2.5. *If $Y = z_0 + X$ where $z_0 \in \mathbf{C}$ and X is a compact subset of $\mathbf{C}$, then Y is compact,*

(i) $t_n(z, Y) = t_n(z - z_0, X)$, $z \in Y$, *and*

(ii) $\|t_n(z, Y)\|_Y = \|t_n(z, X)\|_X$.

Furthermore, equality holds in (i) *when the t_n's are regarded as polynomials. We call any transformation of the form $x \to z_0 + x$ a translation.*

PROOF. The proof is obvious from the definition if card $X < n$; so we assume it is not. Then clearly $t_n(z - z_0, X)$ is monic and has degree n. Suppose that t is an element of P_n. Then by definition of $t_n(z, X)$ we have $\|t(z + z_0)\|_X \geqslant \|t_n(z, X)\|_X$. This gives

$$\|t(z)\|_Y = \|t((z - z_0) + z_0)\|_Y \geqslant \|t_n(z - z_0, X)\|_Y.$$

Thus $t_n(z - z_0, X)$ and $t_n(z, Y)$ are identical as functions on Y, by Proposition 2.3. They are also equal as polynomials since we have assumed that card $X \geqslant n$. Part (ii) is immediate. □

PROPOSITION 2.6. *If z_0 is a nonzero complex number and $Y = z_0X$ where X (and therefore Y) is compact, then*

(i) $t_n(z, Y) = z_0^n t_n(z_0^{-1}z, X)$, $z \in Y$, *and*

(ii) $\|t_n(z, Y)\|_Y = |z_0|^n \|t_n(z, X)\|_X$.

Furthermore, equality holds in (i) *when the t_n's are considered as polynomials. We call any transformation of the form $x \to z_0x$ a dilation.*

PROOF. If card $X < n$, this is obvious so we suppose not. Then it is clear that $z_0^n t_n(z_0^{-1}z, X)$ is monic and of degree n. Suppose that t is an element of P_n. Then by definition of $t_n(z, X)$ we have $|z_0|^{-n}\|t(z_0z)\|_X \geqslant \|t_n(z, X)\|_X$, which gives

$$\|t(z)\|_Y = \|t(z_0(z_0^{-1}z))\|_Y \geqslant |z_0|^n \|t_n(z_0^{-1}z, X)\|_Y.$$

Thus $z_0^n t_n(z_0^{-1}z, X)$ and $t_n(z, Y)$ are equal as functions on Y, by Proposition 2.3. As a result they are equal as polynomials since we have assumed card $X \geqslant n$. Part (ii) is immediate. □

The following is useful in deducing the real case of the approximation problem from the results in the complex case.

PROPOSITION 2.7. *If $X \subset \mathbf{R}$ then $t_n(z, X)$ has real coefficients for each n.*

PROOF. Let $z \in X$. Then $(\operatorname{Re} t_n)(z) = \operatorname{Re}(t_n(z))$ and $(\operatorname{Im} t_n)(z) = \operatorname{Im}(t_n(z))$ so that

$$\|\operatorname{Re} t_n\|_X \leqslant \|\operatorname{Re} t_n + i \operatorname{Im} t_n\|_X = \|t_n\|_X.$$

The result is obvious from the definition of t_n if card $X < n$; so we can assume not. Thus $\operatorname{Re} t_n$ is monic and has degree n. By Proposition 2.3, $t_n = \operatorname{Re} t_n$ as functions on X. As a result they are equal as polynomials since we have assumed card $X \geqslant n$. □

The Čebyšev polynomials which we have defined will now be used to define the transfinite diameter of an arbitrary compact subset of the plane. The definition is based on the following proposition.

PROPOSITION 2.8. *If X is any compact subset of $\mathbf{C}$, then the sequence $\|t_n(z, X)\|_X^{1/n}$, $1 \leqslant n < \infty$, tends to a finite limit.*

PROOF (HILLE [62, Theorem 16.1.2, p. 266]). First note that the sequence in question is bounded, since, by the compactness of X, there is a real positive number r such that X is contained in the disk of radius r centered at the origin. As a result we have $\|t_n\|^{1/n} \leqslant \|z^n\|^{1/n} \leqslant (r^n)^{1/n} = r$. To show that the sequence has a finite limit, we need only show that $\limsup \|t_n\|^{1/n} \leqslant \liminf \|t_n\|^{1/n}$. Let $\varepsilon > 0$ and $\alpha = \liminf \|t_n\|^{1/n}$. There is an integer N such that $\|t_N\|^{1/N} < \alpha + \varepsilon$ which implies $\|t_N\| < (\alpha + \varepsilon)^N$. For any integer $m \geqslant 1$ and integer k such that $0 \leqslant k < N$, the polynomial $z^k t_N^m$ is monic and has degree $mN + k$. Consequently

$$\|t_{mN+k}\| \leqslant \|z^k t_N^m\| < M(\alpha + \varepsilon)^{mN+k},$$

where M is chosen so that $|z| < M^{1/k}(\alpha + \varepsilon)$ for $z \in X$, $1 \leqslant k < N$ and $1 < M$. Then $\|t_{mN+k}\|^{1/(mN+k)} < M^{1/(mN+k)}(\alpha + \varepsilon)$ for $0 \leqslant k < N$. Letting m tend to infinity, we obtain $\limsup \|t_n\|^{1/n} \leqslant \alpha + \varepsilon$. But ε is any positive number; hence

$$\limsup \|t_n\|^{1/n} \leqslant \alpha = \liminf \|t_n\|^{1/n}. \quad \square$$

DEFINITION 2.9. If X is any compact subset of $\mathbf{C}$, we define the *transfinite diameter* of X to be the finite nonnegative number

$$d(X) = \lim_{n\to\infty} \|t_n(z, X)\|^{1/n}.$$

Note that any finite X has transfinite diameter 0 since, for $n > \operatorname{card} X$, $t_n(z, X)$ is identically zero on X.

We note in passing that the definition of $d(X)$ as we have given it is that of the Čebyšev constant of X. The transfinite diameter of a compact subset X of $\mathbf{C}$ was originally defined by Fekete [23] as follows. Let

$$\delta_n(x) = \max\Big(\prod_{1 \leqslant i < j \leqslant n} |z_i - z_j| \Big)^{2/n(n-1)}$$

where the maximum is taken over all n-tuples of elements of X. It is not difficult to see that the $\delta_n(X)$'s form a decreasing sequence bounded below by zero, hence, a convergent sequence. Fekete [23] set $\delta(X) = \lim_{n\to\infty} \delta_n(X)$ and proved that $\delta(X)$ is equal to the Čebyšev constant $d(X)$, as defined in Definition 2.9. It is interesting to note that two other completely different definitions also give rise to the same number $d(X)$; namely, the logarithmic capacity and the exterior mapping radius. (See Tsuji [59].)

Our interest in the concept of the transfinite diameter of X is due to the fact that if it is not less than unity then the uniform approximation problem is trivial. This derives from the following more general fact which is useful in L_p approximation also. Recall that R is any discrete subring of $\mathbf{C}$; hence $|a| \geqslant 1$ whenever $0 \neq a \in R$.

PROPOSITION 2.10. *Let* $(\mathfrak{X}, \|\cdot\|^*)$ *be any normed linear space containing the integral polynomials* $R[z]$, $f \in \mathfrak{X} \setminus R[z]$, *and suppose that* f *is approximable by* *elements of* $R[z]$. *Then there exist monic elements of* $\mathbf{C}[z]$ *with arbitrarily small* $\|\cdot\|^*$ *norms.*

PROOF. Let $\varepsilon > 0$. Since $f \notin R[z]$ there exist distinct elements q, $\tilde{q}$ of $R[z]$ with $\|q - f\|^* < \varepsilon/2 > \|\tilde{q} - f\|^*$. Then

$$\|q - \tilde{q}\|^* = \|q - f + f - \tilde{q}\|^* \leqslant \|q - f\|^* + \|\tilde{q} - f\|^* < \varepsilon.$$

Since q and $\tilde{q}$ are distinct, $q - \tilde{q}$ is not the zero polynomial; hence we can let c be the leading coefficient of $q - \tilde{q}$. Since c is a nonzero element of R, $|c| \geqslant 1$. Thus $\|(q - \tilde{q})/c\|^* < \varepsilon$ and $(q - \tilde{q})/c$ is obviously monic. □

THEOREM 2.11. *Let* X *be a compact subset of* $\mathbf{C}$ *and* R *a discrete subring of* $\mathbf{C}$. *Then*

(i) $d(X) \geqslant 1$ *if and only if every monic polynomial* p *in* $\mathbf{C}[z]$ *satisfies* $\|p\|_X \geqslant 1$, *and*

(ii) *if* $d(X) \geqslant 1$ *then no function on* X *can be uniformly approximated by elements of* $R[z]$ *unless it is identical to an element of* $R[z]$.

PROOF. If $d(X) < 1$ then for some positive integer n we have $\|t_n\|_X^{1/n} < 1$; hence $\|t_n\| < 1$. Conversely, let p be a monic polynomial with $\|p\| = \delta < 1$ and degree k. Then by definition of t_{mk}, $\|t_{mk}\| \leqslant \|p^m\| = \delta^m$ for all positive integers m; hence $\|t_{mk}\|^{1/mk} \leqslant \delta^{1/k} < 1$. This shows that $d(X) = \lim_{n\to\infty}\|t_n\|^{1/n} < 1$ and we have established (i).

Part (ii) is an obvious consequence of Proposition 2.10 and part (i). □

As an example of the use of Proposition 2.10, we show that on an interval of length four or more the L_p approximation problem is trivial. It is also trivial in this case for uniform approximation as we see by taking $p = \infty$. We will see this again in Example 2.15.

THEOREM 2.12. *If* $[a, b]$ *is an interval of length four or more and* $1 \leqslant p \leqslant \infty$, *then no element of* $L_p[a, b] \setminus R[z]$ *can be approximated in* L_p *norm by elements of* $R[z]$.

PROOF. By Proposition 2.10 it suffices to show that there do not exist monic polynomials with arbitrarily small L_p norms on $[a, b]$. Suppose for the moment that on $[-1, 1]$ the monic polynomial m_n of degree $\leqslant n$ and least L_1 norm satisfies $\|m_n\|_1 = 2^{1-n}$. By the usual linear change of variables, it is easy to see that, on $[a, b]$, $\|\tilde{m}_n\|_1 = (b - a)((b - a)/4)^n$ where $\tilde{m}_n$ is the monic polynomial of least L_1 norm on $[a, b]$. For any fixed p, $1 \leqslant p < \infty$, and monic polynomial p_n of degree $\leqslant n$, we have by Hölder's inequality ($p^{-1} + q^{-1} = 1$)

$$\begin{aligned}\|p_n\|_1 &= \int_a^b |p_n(x)|\,dx \leqslant \left(\int_a^b |p_n(x)|^p\,dx\right)^{1/p}\left(\int_a^b 1^q\,dx\right)^{1/q}\\ &= \|p_n\|_p(b - a)^{1/q};\end{aligned}$$

hence

$$\|p_n\|_p \geqslant (b-a)^{-1/q}\|p_n\|_1 \geqslant (b-a)^{1-1/q}((b-a)/4)^n.$$

The same inequalities are valid with the usual modification for $p = \infty$ and the theorem follows.

It remains to show that $\|m_n\|_1 = 2^{1-n}$. This follows from the explicit representation

$$m_n(x) = \frac{1}{2^n}\frac{\sin\left[(n+1)\cos^{-1}x\right]}{\sqrt{1-x^2}}$$

which we proceed to establish. To see that m_n is a monic polynomial of degree n note that it is the product of $(1/(n+1))$ and the derivative of the Čebyšev polynomial $2^{-n}\cos[(n+1)\cos^{-1}x]$ and the latter is a monic polynomial of degree $n+1$ (cf. Lorentz [**66**, §2.7]). It remains to show that m_n has the least L_1 norm among those of all monic polynomials of degree n. We first establish that

$$\int_{-1}^{1} \tilde{p}_{n-1}(x)\operatorname{sgn} m_n(x)\,dx = 0 \tag{$*$}$$

for any polynomial $\tilde{p}_{n-1}$ of degree at most $n-1$. This will follow from

$$\int_{-1}^{1} x^k \operatorname{sgn} \sin\left[(n+1)\cos^{-1} x\right] dx = 0, \qquad 0 \leqslant k < n.$$

Substituting $x = \cos t$ this becomes

$$\int_0^{\pi} \sin t \cos^k t \operatorname{sgn} \sin\left[(n+1)t\right] dt = 0, \qquad 0 \leqslant k < n.$$

The factor $\sin t \cos^k t$ is a trigonometric sine polynomial of order $k+1$ (namely: $2^{-1}(\sin[(k+1)x] - \sin[(k-1)x])$); hence it is enough to establish that

$$\int_0^{\pi} \sin kt \operatorname{sgn} \sin\left[(n+1)t\right] dt = 0, \qquad 0 \leqslant k \leqslant n.$$

However this follows from the identity

$$\int_0^{2\pi} e^{imx} \operatorname{sgn} \sin\left[(n+1)x\right] dx = -e^{im\pi/(n+1)}\int_0^{2\pi} e^{imx} \operatorname{sgn} \sin\left[(n+1)x\right] dx$$

which can be deduced by substituting $x + \pi/(n+1)$ for x in the first integral. The minimality of the norm $\|m_n\|_1$ now follows since for any monic polynomial p_n of degree n we have, with $r_{n-1}(x) = x^n - m_n(x)$,

$$\begin{aligned}\int_{-1}^{1} |p_n(x)|\,dx &= \int_{-1}^{1} |x^n - s_{n-1}(x)|\,dx\\ &\geqslant \int_{-1}^{1} (x^n - s_{n-1}(x))\operatorname{sgn} m_n(x)\,dx\\ &= \int_{-1}^{1} x^n \operatorname{sgn} m_n(x)\,dx - \int_{-1}^{1} s_{n-1}(x)\operatorname{sgn} m_n(x)\,dx\end{aligned}$$

$$= \int_{-1}^{1} x^n \operatorname{sgn} m_n(x)\, dx$$

$$= \int_{-1}^{1} x^n \operatorname{sgn} m_n(x)\, dx - \int_{-1}^{1} r_{n-1}(x) \operatorname{sgn} m_n(x)\, dx$$

$$= \int_{-1}^{1} (x^n - r_{n-1}(x)) \operatorname{sgn} m_n(x)\, dx$$

$$= \int_{-1}^{1} |m_n(x)|\, dx = \|m_n\|_1.$$

In case $p = \infty$ the result also follows from Theorem 2.11 and the fact that if x is an interval, then $d(X)$ is one-fourth of its length (Example 2.15). □

We next give some examples of the calculation of Čebyšev polynomials and transfinite diameters.

PROPOSITION 2.13. *For any complex number z_0 and any compact subset X of* **C** *we have*

(i) $d(z_0 + X) = d(X)$ *and*

(ii) $d(z_0X) = |z_0|d(X)$.

PROOF. Proposition 2.5(ii) shows that $\|t_n(z, z_0 + X)\|_{z_0+X} = \|t_n(z, X)\|_X$ for each n and the conclusion (i) is immediate. To prove (ii), first suppose that z_0 is 0. Then

$$d(z_0X) = d(\{0\}) = 0 \cdot d(X) = |z_0|d(X).$$

On the other hand, suppose z_0 is nonzero. Then 2.6(ii) applies to give

$$\left(\|t_n(z, z_0X)\|_{z_0X}\right)^{1/n} = |z_0|\left(\|t_n(z, X)\|_X\right)^{1/n}.$$

Part (ii) is now immediate. □

EXAMPLE 2.14. We now determine the Čebyšev polynomials and transfinite diameters of all circles $\{z \in \mathbf{C}: |z - z_0| = \alpha\}$ $(\alpha > 0)$ and disks $\{z \in \mathbf{C}: |z - z_0| \leq \alpha\}$ $(\alpha > 0)$. By the maximum modulus principle, the supremum norm of a polynomial over a circle is equal to the supremum norm of that polynomial over the circle, together with any subset of its interior. Thus we will have determined the Čebyšev polynomials for the latter, and in particular for all closed disks, once we have determined them for circles.

Let **T** denote the unit circle in the complex plane. Clearly any circle can be obtained from **T** by a dilation followed by a translation; so by Propositions 2.5 and 2.6, it suffices to find the Čebyšev polynomials for **T**.

We claim that $t_n(z, \mathbf{T}) = z^n$ for all nonnegative integers n. Suppose that $t \in P_n$ and $\|t\|_{\mathbf{T}} < \|z^n\|_{\mathbf{T}}$. Then $|-t(z)| < |z^n|$ for all $z \in \mathbf{T}$, and by Rouché's theorem, $z^n - t(z)$ has the same number of zeros inside **T** as z^n has. That is, $z^n - t(z)$ has n zeros. But, since t is monic by hypothesis, $\deg(z^n - t(z))$ is less than n, which implies that $t(z) = z^n$.

If X is a circle in the complex plane of radius r and center z_0, we have

$X = z_0 + r\mathbf{T}$, and then $d(X) = d(z_0 + r\mathbf{T}) = d(r\mathbf{T}) = rd(\mathbf{T}) = r$ by Proposition 2.13. Thus the transfinite diameter of a circle is equal to its radius. By the remarks at the beginning of this example, the same is true of a disk.

EXAMPLE 2.15. Here we determine the Čebyšev polynomials and transfinite diameter for any line segment in the plane. Let

$$[\alpha, \beta] = \{\lambda\alpha + (1 - \lambda)\beta : 0 \leqslant \lambda \leqslant 1\}.$$

Then clearly

$$[\alpha, \beta] = \frac{\alpha + \beta}{2} + \frac{|\alpha - \beta|}{2}[-1, 1], \tag{$*$}$$

so by Propositions 2.5, 2.6, and 2.13, it suffices to determine the Čebyšev polynomials and transfinite diameter for $[-1, 1]$. By Proposition 2.7 the Čebyšev polynomials for $[-1, 1]$ have real coefficients. Thus, for a given nonnegative integer n, $t_n(z, [-1, 1])$ is just the monic polynomial with degree n and real coefficients whose supremum norm on $[-1, 1]$ is a minimum. It is well known (Lorentz [**66**, Theorem 11, Chapter 1]) that $t_n(x, [-1, 1]) = 2^{1-n}\cos(n(\cos^{-1} x))$, $-1 \leqslant x \leqslant 1$, and from this we see that

$$d([-1, 1]) = \lim_{n\to\infty} 2^{(1-n)/n} = \tfrac{1}{2}.$$

From ($*$) and Proposition 2.13, we have

$$\begin{aligned} d([\alpha, \beta]) &= d\left(\frac{\alpha + \beta}{2} + \frac{|\alpha - \beta|}{2}[-1, 1]\right) = d\left(\frac{|\alpha - \beta|}{2}[-1, 1]\right) \\ &= \frac{|\alpha - \beta|}{2} d([-1, 1]) = \frac{|\alpha - \beta|}{4}. \end{aligned}$$

Thus the transfinite diameter of any line segment is one-fourth of its length.

EXAMPLE 2.16. For explicit calculations of the transfinite diameters of more compact subsets of the plane, circular arcs for example, see Tsuji [**59**].

In the adelic case we are concerned with approximation on sets of the form X_v where X_v is a compact subset of $\tilde{K}_v$, the algebraic closure of the metric completion under v of K. The valuation v has a unique extension to the completion K_v and this in turn has a unique extension to $\tilde{K}_v$. The first extension is unique since it is continuous on K_v and K is dense in K_v. The second extension is unique since $\tilde{K}_v$ can be written as a union of finite extensions of K_v, and extensions of valuations to finite extensions of complete valuated fields are unique (O'Meara [**63**, 14 : 1]). The uniform norm $\|\cdot\|_{X_v}$ for any bounded $\tilde{K}_v$-valued function f_v on X_v is then defined by $\|f_v\|_{X_v} = \sup_{x\in X_v}|f_v(x)|_v$. We extend the concept of transfinite diameter to the present setting as follows.

DEFINITION 2.17. Let $\delta_n(X_v) = \inf\|p\|_{X_v}^{1/n}$ where the infimum is taken over all monic polynomials p in $\tilde{K}_v[x]$ of degree n. Then $\lim_{n\to\infty} \delta_n(X_v)$ exists by essentially the same argument as in Proposition 2.8, and we set $d(X_v) = \lim_{n\to\infty} \delta_n(X_v)$. We note that if $K_v = \mathbf{R}$, $d(X_v)$ is the same as in Definition 2.9 whereas if $K_v = \mathbf{C}$, we obtain the square of the transfinite diameter in Definition 2.9. This is because in the case $K_v = \mathbf{C}$ we have taken $|\cdot|_v$ to be the square of the usual absolute value of $\mathbf{C}$. The squaring of the transfinite diameter does not affect the truth of the criterion $d(X) < 1$ which is all that is used in the sequel. These two cases, $K_v = \mathbf{R}$ and $K_v = \mathbf{C}$, exhaust the possibilities when v is Archimedean (O'Meara [**63**, §12]). Thus when v is Archimedean, the results in Examples 2.14, 2.15, and 2.16 can be used to determine $d(X_v)$ explicitly. In the case of a non-Archimedean valuation v, we have the following (David Cantor [**69**]).

EXAMPLE 2.18. Let K_v be a non-Archimedean completion of the algebraic number field K. Then K_v is a finite algebraic extension of a field of p-adic numbers $\mathbf{Q}_p$. Let f be the residue class degree of K_v over $\mathbf{Q}_p$. If $a \in K_v$ and $S_a = \{x \in K_v : |x|_v \leqslant |a|_v\}$ then $d(S_a) = |a|_v p^{f/(1-p^f)}$.

PROOF. In the present case the result analogous to Proposition 2.13(ii) holds and is proved in essentially the same way; hence we can assume $a = 1$. Since the non-Archimedean valuation v is discrete, we can choose $\pi \in K_v$ such that $|\pi|_v < 1$ and $|\pi|$ generates the group $|K^*|_v$ (O'Meara [**63**, §16]), where K^* denotes the nonzero elements of K. Since $|p|_v$ generates the value group $|\mathbf{Q}_p|_v$, we have $|\pi|_v^e = |p|_v$ where e is the ramification index $[|K_v|_v : |\mathbf{Q}_p|_v]$. But we have normalized $|\cdot|_v$ such that $|p|_v = p^{-n}$ where n is the local degree $[K_v : \mathbf{Q}_p]$. Since the fields are complete we have $n = ef$ (O'Meara [**63**, §16]) and it follows that $|\pi|_v = p^{-f}$.

In Pólya [**19**, Satz IV] it is shown that if P is a polynomial, all of whose coefficients are integers (equivalently, have values $\leqslant 1$), at least one of them has the value 1, and $\|P\|_{S_1} \leqslant |\pi|^\alpha$ then

$$\alpha \leqslant \sum_{k=1}^{\infty} \left[\frac{m}{p^{fk}} \right] = \psi(m)$$

where m is the degree of P and the equality defines $\psi(m)$. Furthermore he shows that there exist monic polynomials P of every degree such that $\|P\|_{S_1} = |\pi|^{\psi(m)}$. Thus by definition

$$\delta_m(S_1) = \left(\inf_P \|P\|_{S_1}\right)^{1/m} = p^{-f\psi(m)/m}.$$

Also notice that $[m/p^{fk}] = 0$ whenever $(m/p^{fk}) < 1$, i.e., whenever $k > \log m/(f \log p) = h(m)$, where the equality serves to define $h(m)$. Hence

$$\sum_{k=1}^{h(m)} \left(\frac{m}{p^{fk}} - 1 \right) \leqslant \sum_{k=1}^{\infty} \left[\frac{m}{p^{fk}} \right] \leqslant \sum_{k=1}^{\infty} \frac{m}{p^{fk}} = m \sum_{k=1}^{\infty} \left(\frac{1}{p^f} \right)^k = \frac{m}{p^f - 1},$$

and we have

$$\sum_{k=1}^{h(m)} \left(\frac{1}{p^f}\right)^k - \frac{\log m}{mf \log p} \leqslant \frac{\psi(m)}{m} \leqslant \frac{1}{p^f - 1}.$$

Since $h(m) \to \infty$ and $(\log m / mf \log p) \to 0$ as $m \to \infty$, we have $\psi(m)/m \to 1/(p^f - 1)$. It follows that

$$d(S_1) = \lim_{m \to \infty} \delta_m(S_1) = p^{f/(1-p^f)}. \quad \square$$

CHAPTER 3

ALGEBRAIC KERNELS

In this chapter we introduce the concept of the algebraic kernel which is due to M. Fekete [**54**]. We will see in the following chapter that for a large class of compact sets X the approximability of a continuous function on X by integral polynomials is determined by the possibility of interpolating it on a finite subset, the algebraic kernel of X. This seems to have been noticed first by Fekete [**54**]. We establish the necessity of the interpolation conditions in this chapter.

DEFINITION 3.1. Let R be any subring of $\mathbf{C}$ and f a complex valued function on a subset X of $\mathbf{C}$. We say that f is *R-interpolable* on a subset S of X if there exists p in $R[z]$ such that $p(z) = f(z)$ for all z in S. We say that f is *R-approximable* on X if for every $\varepsilon > 0$ there exists a p in $R[z]$ with $\|f - p\|_X < \varepsilon$.

DEFINITION 3.2. If R is any subring of $\mathbf{C}$ and X a compact subset of $\mathbf{C}$, we define $B(X, R) = \{p \in R[z] : \|p\|_X < 1\}$.

The set $B(X, R)$ (or simply $B(X)$, if R is understood) can be thought of as the open unit ball of the normed ring $R[z]$, hence the notation.

We note that in Theorem 2.12 we have proved something stronger than stated. In fact we see that if R is a discrete subring of $\mathbf{C}$ and X is a compact subset of $\mathbf{C}$ with $d(X) \geqslant 1$, then $R[z]$ is a discrete and therefore closed subring of $C(X)$. Indeed, we can prove in this case that $B(X, R) = \{0\}$ as follows. If $q \in B(X, R)$ and q is not identically zero on X, then we can divide by its leading coefficient (whose modulus is at least 1 by the remarks following Definition 1.1) to obtain a monic polynomial p with $0 < \|p\|_X < 1$ and derive a contradiction to $d(X) \geqslant 1$ as in the proof of Theorem 2.11. Now, by Hewitt and Ross [**63**, (5.10)], since $R[z]$ is a discrete (additive) subgroup of $C(X)$, it is closed in $C(X)$.

DEFINITION 3.3. For any subring R of $\mathbf{C}$ and compact subset X of $\mathbf{C}$ we define $J(X, R) = \{z \in X\colon\ p(z) = 0 \text{ for all } p \text{ in } B(X, R)\}$. When no confusion is possible, we write $J(X)$ or simply J for $J(X, R)$.

If A is a discrete subring of C with rank 2, then by Proposition 1.10 there exists exactly one imaginary quadratic field L such that $A \subset I_L$. With this fact in mind we make the following definition.

DEFINITION 3.4. Let X be a subset of $\mathbf{C}$ and L the unique imaginary quadratic field such that $A \subset I_L$. We define $J_0(X, A)$ to be the union of the complete sets of conjugates integral over I_L which are entirely contained in X.

We note that in the case $A = I_L$ and X compact, the subset $J_0(X, I_L)$ of X was called the "algebraic kernel" of X with respect to the field L (Fekete [**54**, p. 1338]).

PROPOSITION 3.5. *If X is a compact subset of* $\mathbf{C}$, *then*

$$J_0(X, A) \subset J(X, A).$$

PROOF. Let $\tilde{z} \in J_0(X, A)$ and $A \subset I_L$. Then there is a monic $q \in I_L[z]$ such that $q(\tilde{z}) = 0$ and $Z_q \subset X$ where Z_q denotes the zeros of q. Let $\tilde{q} \in B(X, A)$. We will be done once we show that $\tilde{q}(\tilde{z}) = 0$ which is a consequence of the following. □

LEMMA 3.6. *Let R be a discrete subring of* $\mathbf{C}$, *q and $\tilde{q}$ elements of $R[z]$, q monic, and $\|\tilde{q}\|_{Z_q} < 1$. Then $Z_q \subset Z_{\tilde{q}}$. Thus every element of $B(X, A)$ is identically equal to zero on $J_0(X, A)$.*

PROOF. Let $z_1 \in Z_q$ and let L be an imaginary quadratic field such that $R \subset I_L$ (Proposition 1.9). Then the conjugates over L of z_1, say $z_1, \ldots, z_n$, are also roots of q. Thus, by hypothesis, if we set $\underline{a} = \prod_{j=1}^{n} \tilde{q}(z_j)$, then

$$|\underline{a}| = \prod_{j=1}^{n} |\tilde{q}(z_j)| < 1.$$

Clearly $\underline{a}$ is a symmetric polynomial in the z_j's with coefficients in I_L, hence can be written as a polynomial in the elementary symmetric polynomials of $z_1, \ldots, z_n$ with coefficients in I_L (Jacobsen [**51**, Theorem 9, p. 109]). But the elementary symmetric polynomials in $z_1, \ldots, z_n$ are the coefficients of their minimal polynomial over L. Since $q \in I_L[z]$ and q is monic, z_1 is integral over L; hence its minimal polynomial has coefficients in I_L (Proposition 1.12). Thus $\underline{a} \in I_L$ and $|\underline{a}| < 1$; hence $\underline{a} = 0$. It follows that $\tilde{q}(z_j) = 0$ for some j, hence for $j = 1$ since the z_j's are conjugates. □

PROPOSITION 3.7. *Let R be a discrete subring of* $\mathbf{C}$ *and X a compact subset of* $\mathbf{C}$. *In order that a function f taking X into* $\mathbf{C}$ *be R-approximable on X, it is necessary that f be R-interpolable on $J(X, R)$; if R has rank 2, then f must be interpolable on $J_0(X, R)$ as well.*

PROOF. Suppose that f is R-approximable on X. That is, that there is a sequence $\{p_n\}$ of polynomials in $R[z]$ which tends uniformly to f on X. Then there is an integer N such that $m, n > N$ implies $\|p_n - p_m\| < 1$. That is, $p_n - p_m$ is an element of $B(X, R)$ and so $p_n - p_m = 0$ on $J(X, R)$. Thus, $m > N$ implies that p_m interpolates f on $J(X, R)$. The interpolability of f on $J_0(X, R)$ now follows from Proposition 3.5. □

We have seen in Theorem 2.11(ii) that if $d(X) > 1$ then the problem we are interested in has a trivial solution. On the other hand if $d(X) < 1$, then the set

$J(X, A)$, hence $J_0(X, A)$, is finite for any A. It follows that our problem is not trivial in this case, at least when X is infinite. We will see this later. First an important result due to Fekete **[23]** must be established.

THEOREM 3.8. *Let A be a discrete subring of* $\mathbf{C}$ *with rank* 2 *and X a compact subset of* $\mathbf{C}$. *If there exists a monic polynomial p in* $\mathbf{C}[z]$ *with* $\|p\|_X < 1$, *then there exists a nonzero polynomial q in $A[z]$ with* $\|q\|_X < 1$. *If A contains* 1 *then q can be taken to be monic.*

PROOF. Let n be the degree of p. Define a sequence (starting with the integer n) of monic polynomials as follows. For $m \geqslant n$, set $m = kn + r$ where $0 \leqslant r < n$ and $k \geqslant 1$. Let

$$p_m(z) = z^r p(z)^k. \tag{1}$$

Note that p_m is a monic polynomial of degree m. Also, if $s = \|p\| < 1$, set $t = s^{1/n}$ $(t \geqslant 0)$ so that $s = t^n$. Clearly $0 \leqslant t < 1$. Next pick a real number M such that $\|z^i\| < sM$, $1 \leqslant i < n$, if $s > 0$, or set $M = 0$ if $s = 0$. Then, writing m as above, we have

$$\|p_m\| \leqslant \|p\|^k \|z^r\| \leqslant s^k sM = t^{nk+n}M \leqslant t^{nk+r}M = t^m M.$$

Now fix a positive integer $j \geqslant n - 1$ such that $Mt^{j+1}(1 + \delta(1 - t)^{-1}) < 3^{-1}$ where δ is a positive number such that for any z in $\mathbf{C}$ there exists an element a of A for which $|z - a| < \delta$ (Proposition 1.2). For each $m > j$ we define a polynomial q_m as follows. Set

$$q_m = \alpha_0 p_m + \alpha_1 p_{m-1} + \cdots + \alpha_{m-j-1} p_{j+1} \tag{2}$$

where the α's are defined as follows. Let α_0 be an element of A closest to 1. Thus each q_m is monic if $1 \in A$. Let β be the coefficient of z^{m-1} in $\alpha_0 p_m$. By the way δ was chosen, there is a β' in A such that $|\beta' - \beta| < \delta$. Then if we set $\alpha_1 = \beta' - \beta$, we have that $p' = \alpha_0 p_m + \alpha_1 p_{m-1}$ has leading coefficient α_0 since the degree of p_{m-1} is less than that of p_m. Also, the coefficient of z^{m-1} in p' is the element $\beta + (\beta' - \beta) = \beta'$ of A. Continuing in this way, we pick α's such that $|\alpha_i| < \delta$ for $1 \leqslant i \leqslant m - j - 1$ and the coefficients of $z^m, \ldots, z^{j+1}$ in q_m are elements of A. We have

$$\begin{aligned} \|q_m\| &\leqslant \sum_{i=0}^{m-j-1} \|\alpha_i p_{m-i}\| \leqslant t^m M + \sum_{i=0}^{m-j-1} \delta t^{m-i} M \\ &\leqslant Mt^{j+1}\left(1 + \delta \sum_{i=0}^{m-j-1} t^i\right) \leqslant Mt^{j+1}(1 + \delta(1 - t)^{-1}) < \tfrac{1}{3}. \end{aligned} \tag{3}$$

Next, if $m > j$, we define the $(j + 1)$-tuple $((a_{m0}), \ldots, (a_{mj}))$ as follows. If a_{mi} is the coefficient of z^i in q_m $(0 \leqslant i \leqslant j)$, then let $[a_{mi}]$ be an element of A closest to a_{mi} and set $(a_{mi}) = a_{mi} - [a_{mi}]$ so that $|(a_{mi})| < \delta$. As m varies, these $(j + 1)$-tuples remain in the product space $(\delta D)^{j+1}$, where D is the closed unit disk of $\mathbf{C}$.

Now if $M' = \max\{\|z^i\|_X\colon 0 \leqslant i \leqslant j\}$, we choose $\varepsilon' > 0$ such that $\varepsilon'(j + 1)M' < \frac{1}{3}$. Then, since $(\delta D)^{j+1}$ is compact in the topology given by the norm

$$|||(z_0, \ldots, z_j)||| = \max_{0 \leqslant i \leqslant j} |z_i|,$$

there exist distinct m_1 and m_2 such that

$$|||((a_{m_1 0}), \ldots, (a_{m_1 j})) - ((a_{m_2 0}), \ldots, (a_{m_2 j}))||| < \varepsilon'.$$

We then have

$$\sum_{i=0}^{j} |(a_{m_1 i}) - (a_{m_2 i})| \, \|z^i\| \leqslant (j+1) \max_{0 \leqslant i \leqslant j} \{ |(a_{m_1 i}) - (a_{m_2 i})| \, \|z^i\| \}$$

$$< (j+1)\varepsilon' M' < \tfrac{1}{3}. \tag{4}$$

We combine these estimates as follows. From (3) we infer that

$$\|q_{m_1} - q_{m_2}\| \leqslant \|q_{m_1}\| + \|q_{m_2}\| < \tfrac{2}{3}. \tag{5}$$

If q'_m denotes q_m with $[a_{m_i}]$ in place of a_{m_i} for $0 \leqslant i \leqslant j$, (4) shows that

$$\|(q_{m_1} - q_{m_2}) - (q'_{m_1} - q'_{m_2})\| = \|(q_{m_1} - q'_{m_1}) - (q_{m_2} - q'_{m_2})\|$$

$$\leqslant \sum_{i=1}^{j} |(a_{m_1 i}) - (a_{m_2 i})| \, \|z^i\| < \tfrac{1}{3}. \tag{6}$$

Combining (5) and (6) we obtain $\|q'_{m_1} - q'_{m_2}\| < 1$. Also $q'_{m_1} - q'_{m_2}$ is a nonzero polynomial because each q_m has degree m and $m_1 \neq m_2$. Thus we can take $q = q'_{m_1} - q'_{m_2}$ in the theorem.

If $1 \in A$ the α_0 in (2) is equal to 1 by construction so that each q_m and therefore q'_m is monic. Hence q is also monic. □

PROPOSITION 3.9. *If $d(X) < 1$ and A is any discrete subring of* **C** *with rank* 2, *then $J(X, A)$ and $J_0(X, A)$ are finite.*

PROOF. Since $d(X) < 1$ there exists by Theorem 2.11(i) a monic polynomial p with $\|p\|_X < 1$. Applying Theorem 3.8 we find a $q \in A[z]$ with $\|q\|_X < 1$ and q is not the zero polynomial. Thus $q \in B(X, A)$; hence, by definition, $J(X, A)$ is a subset of the roots of q, hence finite. Since it is a subset of $J(X, A)$, $J_0(X, A)$ is also finite (Proposition 3.5). □

Suppose for the moment that $J_0(X, A)$ is finite. Then we saw in Proposition 3.7 that a function is A-approximable only if it can be interpolated on $J_0(X, A)$ by elements of $A[z]$. At the points of $J_0(X, A)$ which have the property that they, together with all of their conjugates, lie in X° (the interior of X) a stronger interpolation condition obtains, as follows.

PROPOSITION 3.10. *If f is A-approximable on X, a compact subset of* **C** *with $d(X) < 1$, and m is any positive integer then there is a q in $A[z]$ satisfying*

$$q(z) = f(z), \qquad z \in J_0(X, A),$$

and

$$q^{(\nu)}(z) = f^{(\nu)}(z), \qquad z \in J_0(X^\circ, A), \qquad 1 \leqslant \nu \leqslant m.$$

PROOF. Since $d(X) < 1$ we see from Proposition 3.9 that $J_0(X)$, hence $J_0(X^\circ)$, is finite. Let $\rho > 0$ be small enough that the disks with radius ρ and centers in $J_0(X^\circ)$ are contained in X° and do not intersect themselves. Choose a number $\Delta \geqslant 1$ satisfying $\Delta \geqslant \max\{\nu!\rho^{-\nu}\}_{\nu=0}^{m}$. We claim that if $\tilde{q} \in A[z]$ and $\|\tilde{q}\|_X < \Delta^{-1}$, then

$$\tilde{q}(z) = 0, \qquad z \in J_0(x), \tag{1}$$

and

$$\tilde{q}^{(\nu)}(z) = 0, \qquad 1 \leqslant \nu \leqslant m, \qquad z \in J_0(X^\circ). \tag{2}$$

Indeed, (1) holds by Lemma 3.6. From the Cauchy integral formula we have

$$\tilde{q}^{(\nu)}(z) = \frac{\nu!}{2\pi i}\int_{C_z}\frac{\tilde{q}(\xi)}{(\xi - z)^{\nu+1}}\,d\xi, \qquad 1 \leqslant \nu \leqslant m, \qquad z \in J_0(X^\circ),$$

where C_z is the circle with radius ρ and center z. By standard methods, then, we have $|\tilde{q}^{(\nu)}(z)| \leqslant \nu!\rho^{-\nu}\|\tilde{q}\|_X < 1$. If we now replace the X in Lemma 3.6 with the present X° we obtain (2).

Now let $\{q_n\}$ be a sequence from $A[z]$ which converges to f uniformly on X. Then for some n_0

$$\|f - q_n\|_X \leqslant (2\Delta)^{-1}, \qquad n \geqslant n_0;$$

hence

$$\|q_j - q_k\|_X \leqslant \Delta^{-1}, \qquad j, k \geqslant n_0.$$

By our claim the integral polynomial $\tilde{q} = q_j - q_k$ satisfies (1) and (2), i.e., $q_j(z) = q_k(z)$, $z \in J_0(X)$, and $q_j^{(\nu)}(z) = q_k^{(\nu)}(z)$, $1 \leqslant \nu \leqslant m$, $z \in J_0(X^\circ)$, whenever j and k are $\geqslant n_0$. But uniform convergence of $\{q_n\}$ to f on X implies the pointwise convergence of $\{q_n\}$ to f on X and the pointwise convergence of $\{q_n^{(\nu)}\}$ to $\{f_n^{(\nu)}\}$ on X°; hence, the conclusion of the theorem obtains for $q = q_n$, and any $n \geqslant n_0$. □

In the adelic case the analogue of the algebraic kernel is defined as follows. Let K be the algebraic number field and T a finite set of equivalence classes of valuations containing all Archimedean classes, as before. For any three fields E, F, and G with $G \subset E$ and $G \subset F$ we define $\mathrm{Emb}_G(E, F)$ to be the set of isomorphisms of E onto a subfield of F which leaves each element of G fixed.

DEFINITION 3.11. Let J^0 be the set of all $k \in \tilde{K}$, the algebraic closure of K, which are integral over K^T and satisfy $\sigma k \in X_v$ for all $\sigma \in \mathrm{Emb}_K(\tilde{K}, \tilde{K}_v)$. For $v \in T$ define $J_v^0 = (\mathrm{Emb}_K[\tilde{K}, \tilde{K}_v])J^0$, that is, the set of all σj where $\sigma \in \mathrm{Emb}_K[\tilde{K}, \tilde{K}_v]$ and $j \in J^0$.

There are now analogues of each of the earlier propositions and the theorem of this chapter for the adelic case. These will be discussed in the upcoming chapter on this case.

We conclude this chapter with the calculation of J_0 for some specific subsets of the real line and complex plane. Its calculation for general X seems to be very difficult. Since, by definition, $J_0(X, A) = J_0(X, I_L)$ where L is the unique imaginary quadratic field such that $A \subset I_L$, we can restrict ourselves to the case $A = I_L$ in the following without loss of generality.

It is easy to see that if $z_0 \in I_L$ then $J_0(z_0 + X, I_L) = z_0 + J_0(X, I_L)$ and if z_0 is a unit of I_L (i.e., an element of I_L whose multiplicative inverse also lies in I_L), then $J_0(z_0X, I_L) = z_0J_0(X, I_L)$. Thus the following results have a wider range of applicability than first appears.

Throughout the remainder of the chapter, L will represent a fixed but arbitrary imaginary quadratic field, I_L its ring of integers and d the unique square-free rational integer such that $L = \mathbf{Q}(\sqrt{d})$. Since L is imaginary, d is negative. We denote the unit circle $|z| = 1$ by $\mathbf{T}$.

EXAMPLE 3.12. A simple case can be handled by the observation that if $\{z_1, \ldots, z_n\}$ is a complete set of conjugates integral over I_L, then $z_1 \cdots z_n \in I_L$. Indeed this product is, to within sign, the constant term of the minimal polynomial over L of $\{z_1, \ldots, z_n\}$. From this and Proposition 1.12 it follows that the product is an element of I_L. With this result we can see that if X is a subset of a circle centered at the origin and with a transcendental radius ρ then $J_0(X, I_L) = \varnothing$, any I_L. Indeed if $z_1, \ldots, z_n$ is a complete set of conjugates integral over I_L and lying in X, then $z_1 \cdots z_n \in I_L$ as above; hence $\rho^n = |z_1 \cdots z_n|$ is an algebraic number as can be seen by utilizing the representations for the elements of I_L in Proposition 1.6. Thus ρ is algebraic, a contradiction.

EXAMPLE 3.13. See Ferguson [**70a**]. Here we determine $J_0(X, I_L)$ for subsets of the unit disk $D = \{z: (z) \leqslant 1\}$. First, using the observation in Example 3.12 we prove that for any $X \subset D$

$$J_0(X, I_L) = (X \cap \{0\}) \cup J_0(X \cap \mathbf{T}, I_L).$$

Indeed, it is clear that $(X \cap \{0\}) \subset J_0(X, I_L)$ so it suffices to prove that any nonzero element of $J_0(X, I_L)$ is in $J_0(X \cap \mathbf{T}, I_L)$. Let $\{z_1, \ldots, z_n\}$ be a complete set of conjugates integral over I_L with $z_1 \neq 0$ and $\{z_1, \ldots, z_n\} \subset X$. Then $z_1z_2 \cdots z_n$ is the constant term of the minimal polynomial p of the z_i's over L and so is an element of I_L. Each $|z_i| \leqslant 1$ and if $|z_i| < 1$ for some i, then $|z_1z_2 \cdots z_n| < 1$ which implies $z_1z_2 \cdots z_n = 0$ by the discussion following Definition 1.1. But p is irreducible and so must be simply $p(z) = z$. This contradicts the assumption that $z_1 \neq 0$; hence $|z_i| = 1$ for all i and $z_1 \in J(X \cap \mathbf{T}, I_L)$.

This result reduces the problem to finding $J(X, I_L)$ for subsets of the unit circle. Before proceeding with this we need the following generalization of a result due to Kronecker [**1857**].

Let $\{z_1, \ldots, z_n\}$ be a complete set of conjugates integral over I_L where L is an imaginary quadratic field. If $|z_1| = |z_2| = \cdots = |z_n| = 1$ then the z_i's are roots of unity.

To see this, note that the set $\{z_1, \ldots, z_n, \bar{z}_1, \ldots, \bar{z}_n\}$ is a complete set of conjugates over $\mathbf{Q}$ and each element is integral and has absolute value 1. But according to a classical theorem of Kronecker [**1857**], every element of a complete set of conjugate (over $\mathbf{Q}$) algebraic integers all of whose moduli equal 1 is a root of unity.

A corollary of this result is that an element z in X is in $J_0(X, I_L)$ if and only if it is a root of unity and all of its conjugates over L are also in X. This is the criterion that will be exploited in what follows.

Let Arg z denote the branch of the argument function such that $-\pi < \operatorname{Arg} z \leqslant \pi$ for all z. For any positive integer k let $U_k = \{e^{2\pi ij/k}: (j, k) = 1\}$. That is, U_k is the set of primitive kth roots of unity.

Let α satisfy $0 < \alpha \leqslant \pi$. If $X = \{z: |z| = 1, |\operatorname{Arg} z| \geqslant \alpha\}$ then

$$J_0(X, I_L) = \bigcup_{\substack{n\alpha \leqslant 2\pi \\ n \neq 1}} U_n.$$

In fact, it is well known that if z is a primitive nth root of unity, then its set of conjugates over $\mathbf{Q}$ is simply U_n, and so its set of conjugates over L is a subset of U_n. Thus if $U_n \subset X$ then $U_n \subset J_0(X, I_L)$. But if $n\alpha \leqslant 2\pi$, $n \neq 1$, then $U_n \subset X$. On the other hand, if $z_1 \in J_0(X, I_L)$ then $U_n = \{z_1, \ldots, z_k\} \cup \{\bar{z}_1, \ldots, \bar{z}_k\}$ where the z_i's are the conjugates of z_1 over L. Thus z_1 is conjugate to either $\exp(2\pi i/n)$ or $\exp(-2\pi i/n)$ and this element being in X implies $\alpha \leqslant 2\pi/n$.

In the cases treated so far the set $J_0(X, I_L)$ was independent of L. In general, $J_0(X, I_L)$ will depend on L as the following cases show.

Before proceeding with the next case we need to know explicitly the conjugates over L of a given primitive nth root of unity. We reserve the symbol ζ_n $(n > 0)$ for a primitive nth root of unity. Let Δ be the discriminant of the field extension $L/\mathbf{Q}$. The problem of determining the conjugates of ζ_n over L is essentially that of determining the Galois group of the cyclotomic extension $L(\zeta_n)$ of L or at least how it acts on ζ_n. For the Galois group of an extension E/F we write $\Gamma(E/F)$.

It is essentially known that if L is a quadratic field with discriminant Δ over $\mathbf{Q}$, then $L \subset \mathbf{Q}(\zeta_n)$ if and only if $\Delta | n$. In fact a quadratic field $\mathbf{Q}(\sqrt{\Delta}\,)$ with discriminant Δ is contained in the field $Q(\zeta_{|\Delta|})$ of Δth roots of unity and furthermore

$$\sqrt{\Delta} = \sum_{\substack{a \bmod |\Delta| \\ (a, \Delta) = 1,\, a > 0}} (\Delta/a)\zeta_{|\Delta|}^a$$

(Hasse [**65**, p. 513, 5]). The result follows from this and elementary properties of discriminants and cyclotomic fields. Thus if $\Delta \nmid n$ then $L \not\subset \mathbf{Q}(\zeta_n)$; hence $L \cap \mathbf{Q}(\zeta_n) = \mathbf{Q}$. Now to quote, in part, a standard result of Galois theory (cf. Lang [**65**, VIII, §1, Theorem 4]): if K is a Galois extension of k, F is an arbitrary extension of k, and k and F are subfields of some other field, then the restriction map $\sigma \to \sigma|_K$ gives an isomorphism of $\Gamma(KF/F)$ onto $\Gamma(K/K \cap F)$. If we take $K = \mathbf{Q}(\zeta_n)$, $k = \mathbf{Q}$, and $F = L$, then we see that $\Gamma(L(\zeta_n)/L)$ becomes $\Gamma(\mathbf{Q}(\zeta_n)/\mathbf{Q})$ upon restricting each of its elements to $\mathbf{Q}(\zeta_n)$. Since this latter group is well known and $\zeta_n \in \mathbf{Q}(\zeta_n)$, we have what we need (see Corollary 3.13.2

below). If $\Delta | n$ then $L \subset \mathbf{Q}(\zeta_n)$; hence $L(\zeta_n) = \mathbf{Q}(\zeta_n)$ and the following applies.

THEOREM 3.13.1. *If Δ is the discriminant of the extension L of Q and $\Delta | n$, then*

$$\Gamma(\mathbf{Q}(\zeta_n)/L) = \{\sigma_m: (m, n) = (\Delta/m) = 1\}$$

where (Δ/m) is the Kronecker symbol of Δ over m and σ_m is that automorphism of $\mathbf{Q}(\zeta_n)$ such that $\sigma_m(\zeta_n) = \zeta_n^m$ $(m > 0)$.

PROOF. Let $L = \mathbf{Q}(\sqrt{d})$ where d is a square-free rational integer. Throughout the proof m is any positive integer relatively prime to Δ.

It is well known that $\Gamma(\mathbf{Q}(\zeta_n)/\mathbf{Q}) = \{\sigma_m: (m, n) = 1\}$. Since $\Gamma(\mathbf{Q}(\zeta_n)/L)$ consists of those elements of $\Gamma(\mathbf{Q}(\zeta_n)/\mathbf{Q})$ which leave $\sqrt{d}$ fixed, it suffices to prove that $\sigma_m(\sqrt{d}) = (\Delta/m)\sqrt{d}$. But $\Delta = d$ or $4d$, so $(\Delta/m) = (d/m)$, and it suffices to prove that $\sigma_m(\sqrt{d}) = (d/m)\sqrt{d}$. Since L is imaginary, d is negative and we can set $d = -2^j p_1 \cdots p_l = -2^j\omega$ where $j = 0$ or 1 and the p_i's are distinct odd primes. For any odd prime p let $G(\zeta_p)$ denote the Gaussian sum of pth roots of unity. By Lang [**70**, Theorem 6, p. 76]

$$\prod_{i=1}^{l} G(\zeta_{p_i}) = \prod_i \sqrt{(-1/p_i)p_i} = \sqrt{(-1/\omega)\omega}\,,$$

where the heavy brackets denote Jacobi's symbol. By Rademacher [**64**, (10.6), p. 75]

$$\begin{aligned}\sigma_m\left(\sqrt{(-1/m)\omega}\,\right) &= \sigma_m\left(\prod_{i=1}^{l} G(\zeta_{p_i})\right) = \prod_i \sigma_m\left(G(\zeta_{p_i})\right)\\ &= \prod_i G(\zeta_{p_i}^m) = \prod_i (m/p_i) G(\zeta_{p_i})\\ &= \left(\frac{m}{\omega}\right)\sqrt{(-1/\omega)\omega}\,.\end{aligned}$$

If $j = 1$, then d is even, $\Delta = 4d \equiv 0 \pmod 8$ and $(\Delta, m) = 1$ implies m is odd. Since $\Delta | n$ we have that $\zeta_8 \in Q(\zeta_n)$ and

$$\sigma_m(2^{j/2}) = \left[\sigma_m(\zeta_8 + \zeta_8^{-1})\right]^j = (2/m)^j 2^{j/2}$$

as is easily seen. For $j = 0$ this result is trivial.

We now consider two cases according to whether $(-1/\omega)$ is $+1$ or -1. If $(-1/\omega) = -1$ then since $(-1/\omega) = (-1)^{(\omega-1)/2}$ we have $\omega \equiv -1 \pmod 4$,

$$\sqrt{d} = \sqrt{(-1/\omega)\omega}\; 2^{j/2}$$

and by the previous paragraph

$$\sigma_m(\sqrt{d}\,) = (m/\omega)\sqrt{(-1/\omega)\omega}\;(2/m)^j 2^{j/2}. \tag{1}$$

Note that $(-\omega/m)$ makes sense as a Kronecker symbol even when m is even since $-\omega \equiv 1 \pmod 4$. Then by the reciprocity law for the Kronecker symbol (Hasse [**64**, p. 143]) we have $(-\omega/m) = (m/-\omega) = (m/\omega)$. From (1) then

$$\sigma_m(\sqrt{d}\,) = (-\omega/m)(2/m)^j\sqrt{d} = (-2^j\omega/m)\sqrt{d} = (d/m)\sqrt{d}\,.$$

It remains to suppose $(-1/\omega) = 1$. Since $(-1/\omega) = 1$ we have $\omega \equiv 1 \pmod 4$,

$$\sqrt{d} = i2^{j/2}\sqrt{(-1/\omega)\omega}\ ,$$

and

$$\begin{aligned}\sigma_m(\sqrt{d}\,) &= \sigma_m(i)(2/m)^j 2^{j/2}(m/\omega)\sqrt{(-1/\omega)\omega}\\ &= \sigma_m(i)(m/\omega)(2/m)^j 2^{j/2}\sqrt{(-1/\omega)\omega}\\ &= -i\sigma_m(i)(m/\omega)(2/m)^j\sqrt{d}\ .\end{aligned}$$

Since $i = \zeta_4$, we have

$$\sigma_m(\sqrt{d}\,) = \zeta_4^3\zeta_4^m(m/\omega)(2/m)^j\sqrt{d}\ . \tag{2}$$

Since $\omega \equiv 1 \pmod 4$ we may again apply the reciprocity law for the Kronecker symbol and write $(m/\omega) = (\omega/m)$. Furthermore $\zeta_4^{m+3} = \zeta_4^{m-1} = (-1)^{(m-1)/2} = (-1/m)$. Substituting these two results into (2) gives

$$\begin{aligned}\sigma(\sqrt{d}\,) &= (-1/m)(\omega/m)(2/m)^j\sqrt{d} = (-\omega 2^j/m)\sqrt{d}\\ &= (d/m)\sqrt{d}\ .\ \square\end{aligned}$$

We note in passing that the real case ($d > 0$) is proved just as easily by the same techniques. We omit the details since we do not have need of that case here.

COROLLARY 3.13.2 (FERGUSON [**70a**]). *If ζ_n is a primitive nth root of unity, then the conjugates of ζ_n over L are*

(i) $\{\zeta_n^m: (m, n) = 1\}$ *if* $\Delta \nmid n$, *or*

(ii) $\{\zeta_n^m: (m, n) = (\Delta/m) = 1\}$ *if* $\Delta | n$

where (Δ/m) *is Kronecker's symbol.*

We note for completeness that (Weiss [**63**, 6-1-1]) Δ is given by

$$\Delta = \begin{cases} d & \text{if } d \equiv 1 \pmod 4,\\ 4d & \text{if } d \not\equiv 1 \pmod 4.\end{cases}$$

From definitions it is easy to see that $J_0(\overline{X}, I_L) = \overline{J_0(X, I_L)}$ where the bar denotes complex conjugation. For this reason the hypothesis $\alpha < (2\pi - \beta)$ in the following entails no loss of generality. (Equality would put us in the situation above.)

Suppose $X = \{e^{i\theta}: \alpha \leqslant \theta \leqslant \beta\}$ with $0 < \alpha < \beta < 2\pi$ and $\alpha < (2\pi - \beta)$. Then we have

$$J_0(X, I_L) = \Bigg(\bigcup_{n(2\pi-\beta)<2\pi;\ n>1} U_n\Bigg) \cup \Bigg(\bigcup_n \{e^{2\pi i m/n}: (m, n) = (\Delta/m) = 1\}\Bigg)$$

where the last union ranges over all positive integers n such that $\alpha \leqslant (2\pi/n) < (2\pi - \beta)$, $\Delta | n$, and $\max\{2\pi m/n: (m, n) = (\Delta/m) = 1\} \leqslant \beta$.

We prove the two sets equal by establishing inclusion both ways. First notice that

$$J_0(X, I_L) \supset \left(\bigcup_{n(2\pi-\beta)\leqslant 2\pi;\, n>1} U_n \right)$$

as above since if this union is nonvoid, we have $(2\pi - \beta) \leqslant \pi$ and $X \supset \{z: |z| = 1, |\text{Arg } z| \geqslant (2\pi - \beta)\}$. Also, if $\Delta | n$ then the conjugates over L of $\zeta_n = e^{2\pi i/n}$ are $\{\zeta_n^m: (m, n) = (\Delta/m) = 1\}$ by Theorem 3.13.1. Thus if $\alpha \leqslant (2\pi/n)$ and $\max\{2\pi m/n: (m, n) = (\Delta/m) = 1\} \leqslant \beta$ then $e^{2\pi i/n}$ and all of its conjugates over L are in X, hence in $J(X, I_L)$.

On the other hand, suppose x is an element of $J_0(X, I_L)$. Then as above x is a root of unity and all of its conjugates over L are in X. Thus x is a primitive nth root of unity for some integer $n > 1$. If $\Delta \nmid n$, then the conjugates of x over L form the set U_n of all primitive nth roots of unity. Thus $e^{2\pi i(n-1)/n}$ is in X which implies that $2\pi(n-1)/n \leqslant \beta$ or $(2\pi - \beta) \leqslant (2\pi/n)$; hence x is an element of $\bigcup_{k(2\pi-\beta)\leqslant 2\pi; k>1} U_k$. If $\Delta | n$, then the conjugates of x over L are $\{x^m: (m, n) = (\Delta/m) = 1\}$ by Corollary 3.13.2. We easily see that the conjugates of x over L together with their complex conjugates form the set of all primitive nth roots of unity. Thus x is conjugate over L to $e^{2\pi i/n}$ or $e^{2\pi i(n-1)/n}$. If x is conjugate to $e^{2\pi i(n-1)/n}$, then x is an element of $\bigcup_{k(2\pi-\beta)\leqslant 2\pi} U_k$ as before. If not, then $e^{2\pi i/n}$ is in X and $(2\pi/n) \geqslant \alpha$. If $2\pi \geqslant n(2\pi - \beta)$, then x is an element of $\bigcup_{k(2\pi-\beta)\leqslant 2\pi} U_k$. Thus we can assume that $\alpha \leqslant 2\pi/n \leqslant 2\pi - \beta$. Since x is conjugate over L to $e^{2\pi i/n}$, the conjugates of $e^{2\pi i/n}$ over L are the same as those of x and $e^{2\pi i m/n}$ is in X whenever $(m, n) = (\Delta/m) = 1$. Thus

$$\max\{2\pi m/n: (m, n) = (\Delta/m) = 1\} \leqslant \beta$$

and x is contained in the second union in the result.

For the real case a different definition of J_0 must be made. Here the ring of integers is simply the rational integers $\mathbf{Z}$ which is not of rank 2; hence the set $J_0(X, \mathbf{Z})$ is undefined up to now.

DEFINITION 3.14. For any compact subset X of $\mathbf{R}$ let $J_0(X, \mathbf{Z})$ be the union of all complete sets of conjugates integral over $\mathbf{Z}$ and entirely contained in X.

We will see later in the chapter on the real case that in fact $J_0(X, \mathbf{Z}) = J_0(X, A)$ for any discrete subring A of $\mathbf{C}$ with rank 2 provided $d(X) < 1$ and $X \subset \mathbf{R}$.

EXAMPLE 3.15. See Hewitt and Zuckerman [**59**, §5]. Here we will calculate $J_0(X, Z)$ (hereinafter denoted $J_0(X)$) for subsets X of $[-2, 2]$. The applicability is enlarged when one considers that for any x_0 in $\mathbf{Z}$ we have $J_0(x_0 + X) = x_0 + J_0(X)$ and $J_0(-X) = -J_0(X)$, as is easily seen.

First notice that if x is an algebraic integer all of whose conjugates lie in $[-2, 2]$, then the associated minimal polynomial is monic with integral coefficients (Proposition 1.12), hence by Kronecker [**1857**], x has the form $2 \cos 2\pi j/k$, j and k integers. We can assume $(j, k) = 1$ without loss of generality and also $1 \leqslant j \leqslant k/2$ since $\cos 2\pi j/k = \cos 2\pi(k - j)/k$. Thus we see that for $X \subset [-2, 2]$ we have $J_0(X) \subset \bigcup_{k=1}^{\infty} T_k$ where by definition $T_k = \{2 \cos 2\pi j/k\colon 0 \leqslant j \leqslant k/2 \text{ and } (j, k) = 1\}$.

We next claim that given $x = 2 \cos 2\pi j/k$ where $1 \leqslant j \leqslant k/2$ and $(j, k) = 1$ then x is an algebraic integer and its set of conjugates is T_k. Fix $\zeta_k = e^{2\pi i/k}$. We saw in Example 3.11 that $\Gamma(\mathbf{Q}(\zeta_k), \mathbf{Q})$ is $\{\sigma_j\}_{(j,k)=1 \leqslant j \leqslant k}$. But $\zeta_k + \zeta_k^{-1} \in \mathbf{Q}(\zeta_k)$; hence its conjugates are

$$\zeta_k^j + \zeta_k^{-j} = 2 \cos 2\pi j/k, \qquad (j, k) = 1 \leqslant j \leqslant k. \tag{$*$}$$

This is easily seen to be the same as $2 \cos 2\pi j/k$, $(j, k) = 1 \leqslant j \leqslant k/2$, since

$$\sigma_j(\zeta_k + \zeta_k^{-1}) = \zeta_k^j + \zeta_k^{-j} = \sigma_{k-j}(\zeta_k + \zeta_k^{-1})$$

and $(k - j, k) = (j, k)$. Since ζ_k satisfies the polynomial $x^k - 1$, it is an algebraic integer and by Proposition 1.4 the algebraic integers form a ring. By writing the elements of T_k in the form $(*)$ we see that they are algebraic integers as well.

From the definition it is now clear that $J_0(X)$ is the union of those T_k which are subsets of X. For subintervals of $[-2, 2]$ this can be put in a form which is more useful for computing J_0 as follows. Let the interval be $[a, b]$.

First notice that if $k \geqslant 3$, then the largest element of T_k is $2 \cos 2\pi j/k$ where $j = 1$ and the smallest occurs when

$$\begin{aligned} j &= (k-1)/2 && \text{if } k \text{ is odd,} \\ j &= (k-2)/2 && \text{if } k \equiv 0 \pmod 4\text{, and} \\ j &= (k-4)/2 && \text{if } k \equiv 2 \pmod 2. \end{aligned}$$

We omit the simple proof. Thus

$$J_0([\alpha, \beta]) = ([\alpha, \beta] \cap \{-2\}) \cup ([\alpha, \beta] \cap \{2\}) \cup \Big(\bigcup_k T_k\Big)$$

where the last union is over all $k \geqslant 3$ such that $2 \cos 2\pi/k \leqslant \beta$ and

$$\begin{aligned} \alpha &\leqslant 2 \cos \pi(k-1)/k && \text{if } k \text{ is odd,} \\ \alpha &\leqslant 2 \cos \pi(k-2)/k && \text{if } k \equiv 0 \pmod 4\text{, or} \\ \alpha &\leqslant 2 \cos \pi(k-4)/k && \text{if } k \equiv 2 \pmod 4. \end{aligned}$$

Thus the problem is reduced to the calculation of the sets T_k.

PART II: QUALITATIVE RESULTS

CHAPTER 4

COMPLEX CASE I: VOID INTERIOR

In this chapter we will give a necessary and sufficient condition for functions on certain compact subsets of the complex plane to be approximable by integral polynomials. In Chapter 7 we will consider the same problem for more general compact subsets of $\mathbf{C}$ and $\mathbf{C}^n$.

DEFINITION 4.1. A compact subset X of $\mathbf{C}$ is said to be Lavrent′ev if $\mathbf{C}[z]$ is uniformly dense in $C(X)$.

The terminology stems from the fact that in 1934 Lavrent′ev established the difficult part of the following proposition.

PROPOSITION 4.2. *A compact subset X of $\mathbf{C}$ is Lavrent′ev if and only if it has void interior and connected complement.*

PROOF. If X has void interior and connected complement, then X is Lavrent′ev by Lavrent′ev [**36**, p. 25]. Conversely, if X is Lavrent′ev then the continuous function $z \to \bar{z}$ restricted to X is a uniform limit of polynomials in z. But a uniform limit of such polynomials is holomorphic at every interior point and $z \to \bar{z}$ is holomorphic at no point, so X has void interior. To prove that X has a connected complement we assume that it does not, and derive a contradiction. Let U be a bounded connected component of $\mathbf{C} \setminus X$ and z_0 an element of U. Then the function $f(z) = (z - z_0)^{-1}$ restricted to X is continuous, hence uniformly approximable by polynomials on X. Let $\{p_n\}$ be a sequence of polynomials tending to f on X. It is easy to see that the boundary of U is contained in X; hence by the maximum modulus theorem, $\{p_n\}$ converges uniformly on U as well. Let g be the uniform limit of $\{p_n\}$ on U. Then $(z - z_0)p_n(z) \to 1$ uniformly on X and $(z - z_0)p_n(z) \to (z - z_0)g(z)$ uniformly on U. Both $(z - z_0)g(z)$ and $g(z)$ are holomorphic on U since they are uniform limits of polynomials. Using the maximum modulus principle again, $(z - z_0)g(z) \equiv 1$ on U. Thus $g(z) = 1/(z - z_0)$ on $U \setminus \{z_0\}$ which contradicts the previous conclusion that g is holomorphic on U. $\square$

As in the proof of Proposition 3.8 we can apply Theorem 3.7 to obtain the following.

LEMMA 4.3. *Let A be a discrete subring of* $\mathbf{C}$ *with rank* 2, *and X any compact subset of* $\mathbf{C}$. *If $d(X) < 1$, then there is a nonzero polynomial q in $A[z]$ with $\|q\|_X < 1$. If, furthermore, A contains the identity element, then q can be taken to be monic.*

LEMMA 4.4. *For any polynomials p, q in $\mathbf{C}[z]$ such that $n = \deg q$ is greater than zero, we can write*

$$p = p_0 + p_1 q + \cdots + p_k q^k$$

where each p_i is a polynomial with $\deg p_i < n$ $(0 \leqslant i \leqslant k)$ and $k = [\deg p/n]$.

PROOF. We proceed by induction on $\deg p$. If $\deg p$ is less than n, the result is obvious. Let $\deg p = m$ be greater than or equal to n and suppose that the result holds for all polynomials p with $\deg p < m$. Then we can write

$$p = p'q + p_0 \tag{1}$$

by the division algorithm, where p' and p_0 are polynomials and $\deg p_0 < n$. Then $\deg p'$ is less than $\deg p$, since if $\deg p' = 0$ this is obvious, and if not then $m = \deg p' + \deg q$ which is greater than $\deg p'$ by the hypothesis on q. Thus the induction hypothesis applies to p' and we can write

$$p' = p_1 + p_2 q + \cdots + p_k q^{k-1}. \tag{2}$$

By substituting (2) into (1) we obtain the desired form for p and the induction is complete. □

LEMMA 4.5. *Let q be a monic polynomial in $A[z]$ with $\|q\|_X < 1$ and b in $\mathbf{C}[z]$. Then there exist $[b]$ in $A[z]$ and M not depending on b such that $\|b - [b]\|_X < M$.*

PROOF. Since $n = \deg q$ is at least 1 we can write $b = b_0 + b_1 q + \cdots + b_k q^k$ where each $b_i \in \mathbf{C}[z]$ and $\deg b_i < \deg q$ $(0 \leqslant i \leqslant k)$. For each i let $[b_i]$ be the polynomial obtained from b_i by replacing each coefficient by a nearest element of A. Then with δ as in Proposition 1.2 we have

$$\|b_i - [b_i]\| \leqslant \sum_{j=0}^{n-1} \|\delta z^j\| = M_0, \qquad 0 \leqslant i \leqslant k,$$

where the last equality serves to define M_0. Thus

$$\left\| b - \sum_{i=0}^{k} [b_i] q^i \right\| = \sum_{i=0}^{k} (b_i - [b_i]) q^i$$

$$\leqslant \sum_{i=0}^{k} \|b_i - [b_i]\| \, \|q\|^i < M_0 (1 - \|q\|)^{-1}. \quad \square$$

LEMMA 4.6. *Let X be a compact subset of* $\mathbf{C}$ *and suppose further that*
(i) *X is Lavrent'ev;*
(ii) *$f \in C(X)$;*
(iii) *q is a monic polynomial in $A[z]$ with $\|q\|_X < 1$;*

(iv) *for any* $\varepsilon > 0$ *there is an* r *in* $A[z]$ *such that* $|f(z) - r(z)| < \varepsilon$ *whenever* $q(z) = 0$ *and* $z \in X$.

Then f *is* A*-approximable on* X.

PROOF. Let Z_q be the set of roots of q which lie in X. Let ε be any positive number. By (iv) there is an r in $A[z]$ such that $|f(z) - r(z)| < \varepsilon/4$, $z \in Z_q$. Then by continuity there is, for each α in Z_q, a closed disk D_α with center α and radius ρ_α such that the family $\{D_\alpha\}_{\alpha \in Z_q}$ is pairwise disjoint and $|f(z) - r(z)| < \varepsilon/2$, $z \in D_\alpha \cap X$. Plainly there is a continuous function u mapping X into $[0, 1]$ such that $u(z) = 1$ for z in no D_α and $u(z) = 0$ if for some α in Z_q, z is in the closed disk of radius $\rho_\alpha/2$ centered at α. It is easy to see that

$$\|u(f - r) - (f - r)\| < \varepsilon/2. \tag{1}$$

By Lemma 4.5 there is a positive integer N such that

$$\|bq^N - [b]q^N\| \leqslant \varepsilon/4 \tag{2}$$

for every b in $\mathbf{C}[z]$. Now consider $u(f - r)/q^N$, which is defined to be zero whenever q is zero. It is continuous by construction. Thus by (i), there is an element b in $\mathbf{C}[z]$ such that

$$\|u(f - r)/q^N - b\| < \varepsilon/4.$$

It follows that $\|u(f - r) - bq^N\| < \|q\|^N\varepsilon/4 < \varepsilon/4$. Then, by (2), $\|u(f - r) - [b]q^N\| < \varepsilon/2$ and by (1), $\|(f - r) - [b]q^N\| < \varepsilon$ or $\|f - (r + [b]q^N)\| < \varepsilon$. □

THEOREM 4.7 (FERGUSON **[68a]**). *Let* X *be a Lavrent'ev subset of* $\mathbf{C}$ *with* $d(X) < 1$. *If* f *is a complex valued function on* X, *then the following are equivalent*:

(i) f *is* A*-approximable on* X;

(ii) f *is continuous and* A*-interpolable on* $J_0(X, A)$.

PROOF. From Proposition 3.7 we see that (i) implies (ii). To prove the converse first note that if $f = p$ on $J_0(X, A)$ and $p \in A[z]$, then it suffices to approximate $f - p$. Since $f - p = 0$ on $J_0(X, A)$ we will assume without loss of generality that $f = 0$ on $J_0(X, A)$. Let L be the imaginary quadratic field such that $A \subset I_L$ (Proposition 1.10). By Proposition 1.11 there exists a positive integer m such that $mI_L \subset A$. Thus if $p \in I_L[z]$ and $\|f/m - p\| < \varepsilon/m$, then $\|f - mp\| < \varepsilon$ and $mp \in A[z]$. In view of this we will assume without loss of generality that $A = I_L$.

Thus A, X, and f satisfy (i) and (ii) of Lemma 4.6, and it only remains to show that (iii) and (iv) hold. By Lemma 4.3 there exists a monic q in $A[z]$ with $\|q\|_X < 1$ so part (iii) of Lemma 4.6 is satisfied. Let Z_q denote the zeros of q which lie in X. Write $J_0(X, A)$ as the union of the sets of zeros of a set of monic irreducible polynomials $\{q_1, \ldots, q_s\}$ in $I_L[z]$. Denote the remaining elements of Z_q by $\alpha_1, \ldots, \alpha_k$ so that

$$Z_q = J_0(X, A) \cup \{\alpha_1, \ldots, \alpha_k\}.$$

By definition of $J_0(X, A)$, the α_i's form a set of algebraic numbers which does not contain a complete set of conjugates over L. In view of this, Theorem A.2 (Appendix) can be applied to give $\tilde{q}$ in $A[z]$ such that

$$\left|\tilde{q}(\alpha_i) - \frac{f(\alpha_i)}{q_1 \cdots q_s(\alpha_i)}\right| < \frac{\varepsilon}{|q_1 \cdots q_s(\alpha_i)|}, \qquad 1 \leqslant i \leqslant k.$$

Then $|\tilde{q}q_1 \cdots q_s(z) - f(z)| < \varepsilon$, $z \in Z_q$, and $\tilde{q}q_1 \cdots q_s \in A[z]$ which shows that (iv) is satisfied. □

THEOREM 4.8. *Let X be a Lavrent'ev subset of $\mathbf{C}$ with $d(X) < 1$ and $A = I_L$. Then a continuous complex valued function f on X is A-approximable if and only if its Lagrange interpolating polynomial r on $J_0(X, A)$ is an element of $A[z]$.*

PROOF. By Theorem 4.7 the condition $r \in A[z]$ is sufficient for the A-approximability of f since r interpolates f on $J_0(X, A)$. Conversely, from Theorem 4.7 we know that if f is A-approximable, then there is a p in $A[z]$ which interpolates f on $J_0(X, A)$. Let $q_1, \ldots, q_s$ be as in the proof of Theorem 4.7. Since each q_i is irreducible, it has only simple roots. Thus $\deg q_1 \cdots q_s = \operatorname{card} J_0(X, A)$ which we denote by n. Since $q_1 \cdots q_s$ is a monic polynomial in $A[z]$, we can find w, t in $A[z]$ such that $p = w(q_1 \cdots q_s) + t$, $\deg t < n$, by the division algorithm. Thus $t = p = f$ on $J_0(X, A)$ and then by the uniqueness of Lagrange interpolating polynomials $t = r$ and $r = f$ on $J_0(X, A)$. □

The following results are useful in determining the set $J_0(X, A)$. In particular, we will see that often $J_0(X, A) = J(X, A)$. The method of proof is adapted from D. Cantor [**69**].

THEOREM 4.9. *Let X be a Lavrent'ev subset of $\mathbf{C}$ with $d(X) < 1$. Then $J_0(X, A)$ is finite and there exists $q \in A[z]$, $q \neq 0$, with $J_0(X, A) = Z_q \cap X$ and $\|q\|_X < 1$.*

PROOF. Since $d(X) < 1$ we have from Proposition 3.9 that $J_0(X, A) = J_0$ is finite. Let L be the unique imaginary quadratic field such that $I_L \supset A$ (Proposition 1.10). It suffices to find $\tilde{q}$ in $I_L[z]$ with $\|\tilde{q}\| < 1$ and $J_0 = Z_{\tilde{q}} \cap X$. Indeed, let $m_0 I_L \subset A$ as in Proposition 1.11. Then for a sufficiently large positive integer n, $q = m_0\tilde{q}^n$ will satisfy the conclusions of the present theorem. Consider the product q_0 of the minimal polynomials of the elements of J_0. It is clear from Definition 3.4 of J_0 and Proposition 1.12 that q_0 lies in $I_L[z]$ and $Z_{q_0} = J_0$. Let $K = \{x \in X\colon |q_0(x)| \geqslant \frac{1}{2}\}$ and $a = \min\{1, (2\|q_0\|_K)^{-1}\}$. (If $\|q_0\|_K = 0$ then we can set $q = q_0$ and be done.) Then K and Z_{q_0} are disjoint closed sets; hence there is a continuous function f on X such that

$$f(Z_{q_0}) \subset \{1\}, \qquad f(K) \subset \{a\},$$

and

$$a \leqslant f(x) \leqslant 1 \qquad \text{all } x \in X.$$

Notice that

$$\|fq_0\| \leqslant \tfrac{1}{2}. \tag{1}$$

The function f is interpolable on J_0 by the element 1 in $I_L[z]$; hence by Theorem 4.7 there exists $q_1 \in I_L[z]$ with

$$\|q_1 - f\| < \min\{a, (2\|q_0\|)^{-1}\}. \tag{2}$$

Then $q_1(x) \neq 0$, for all $x \in X$; hence $Z_{q_1q_0} \cap X = Z_{q_0} \cap X = Z_{q_0}$ and

$$\begin{aligned}\|q_1q_0\| &= \|q_1q_0 - fq_0 + fq_0\| \leqslant \|q_1q_0 - fq_0\| + \|fq_0\| \\ &\leqslant \|q_1 - f\| \, \|q_0\| + \|fq_0\| < 1\end{aligned}$$

by (1) and (2). Now take $q = q_1q_0$. □

THEOREM 4.10. *Let X be a Lavrent'ev subset of* **C** *with $d(X) < 1$. Then $J_0(X, A) = J(X, A)$.*

PROOF. From Proposition 3.5, $J_0 = J_0(X, A) \subset J(X, A) = J$. The reverse inclusion is proved as follows. Let q be as in Theorem 4.9 so that $Z_q \cap X = J_0(X, A)$. Set $Y = Z_q \setminus X$. Then $\|q\|_{X \cup Y} < 1$. If a is the leading coefficient of q then $|a| \geqslant 1$ by the remarks following Definition 1.1; hence $\|a^{-1}q\|_{X \cup Y} < 1$. But $a^{-1}q$ is a monic polynomial and we see as in the proof of Theorem 2.11 that $d(X \cup Y) < 1$. Clearly Y is a finite set, so from Proposition 4.2 we see that $X \cup Y$ is Lavrent'ev. After multiplying the polynomial q_0 in the proof of Theorem 4.9 by the m_0 from Proposition 1.11, we get $q_1 \in A[z]$ such that $Z_{q_1} = J_0$. Define a function f on $X \cup Y$ by $f = q_1$ on Y and $f = 0$ on X. Then $f - q_1$ is continuous on $X \cup Y$ and A-interpolable by 0 on Z_q. From the definition of $J(X \cup Y, A)$ and Proposition 3.5

$$Z_q \supset J(X \cup Y, A) \supset J_0(X \cup Y, A);$$

hence $f - q_1$ is A-interpolable on $J_0(X \cup Y, A)$. By Theorem 4.7, $f - q_1$ is A-approximable on $X \cup Y$. That is, we can find a q_2 in $A[z]$ such that

$$\|f - (q_1 + q_2)\|_{X \cup Y} < \min\left\{\min_{x \in Y} |f(x)|, 1\right\}.$$

From this we conclude that the polynomial $q_3 = q_1 + q_2$ satisfies $q_3(x) \neq 0$, $x \in Y$, and $\|q_3\|_X < 1$. It is clear that $Z_q \cap Z_{q_3} \subset J_0(X, A)$; hence, by definition, $J(X, A) \subset J_0(X, A)$. □

The following is an interesting consequence of the proof of Theorem 4.10.

COROLLARY 4.11. *Let X be a Lavrent'ev subset of* **C** *with $d(X) < 1$. Then there exist q_1 and q_2 in $A[z]$ with $\|q_i\| < 1$, $i = 1, 2$, and $J(X, A) = Z_{q_1} \cap Z_{q_2}$.*

It is not generally possible to find a single integral polynomial with norm less than unity with exactly $J(X, A)$ as its set of zeros as the following shows.

EXAMPLE 4.12 (CANTOR [**69**]). Let

$$X = \left[-\tfrac{1}{2}, \tfrac{1}{2}\right] \cup \left[\tfrac{5}{4}, \tfrac{6}{4}\right] \quad \text{and} \quad A = \mathbf{Z} + i\mathbf{Z}.$$

It is clear from Proposition 4.2 or Weierstrass' theorem that X is Lavrent'ev.

From Example 2.13, $d([-\frac{1}{2}, \frac{3}{2}]) = \frac{1}{2}$; hence $d(X) \leqslant \frac{1}{2}$. Let $q(x) = x(x-1)$. Then $\|q\|_X = \frac{3}{4}$; hence

$$J(X, A) \subset Z_q \cap X = \{0\}.$$

Clearly $0 \in J_0(X, A)$; hence, by Proposition 3.5, $J(X, A) = \{0\}$. There is no polynomial $\tilde{q}$ in $A[z]$ with $J(X, A) = Z_{\tilde{q}}$ and $\|\tilde{q}\| < 1$ since the first implies that $\tilde{q}$ must have the form mz^n for some $m \in A$ and positive integer n but $\|mz^n\|_X \geqslant 1$.

In Theorem 2.10 we showed that if $d(X) \geqslant 1$ then $A[z]$ is already uniformly closed in $C(X)$. The following is a partial converse to that result.

PROPOSITION 4.13. *Let X be a Lavrent'ev subset of* $\mathbf{C}$ *with $d(X) < 1$. Then $A[z]$ is uniformly closed in $C(X)$ if and only if $J(X, A) = X$. In particular, if X is infinite, then $A[z]$ is not uniformly closed in $C(X)$.*

PROOF. Suppose that $J = J(X, A) = X$ and that $f \in C(X)$. By Theorems 4.7 and 4.10, if f is A-approximable on X, it is A-interpolable on $J = X$ so $f \in A[z]$ (considered as functions on X), which shows that $A[z]$ is uniformly closed in $C(X)$.

On the other hand, if $J \neq X$, let z_0 be a point in X but not in J. For any $y \in \mathbf{R}$ we can define a continuous function f_0: $J \cup \{z_0\} \to \mathbf{C}$ by $f_0(J) = \{0\}$ and $f_0(z_0) = y$. It is continuous where defined since $d(X) < 1$ which implies that J is finite (Proposition 3.8) so that the relative topology on $J \cup \{z_0\}$ is discrete. By Tietze's extension theorem there is a continuous extension f of f_0 to all of X. This extension f is obviously A-interpolable on J and so is A-approximable on X by Theorem 4.7. Since y is any real number, this shows that there are uncountably many A-approximable f in $C(X)$. On the other hand, $A[z]$ is countable, since A is, so $A[z] \neq A[z]^-$ where the bar denotes uniform closure in $C(X)$.

The last statement now follows from the fact that $J(X, A)$ is finite whenever $d(X) < 1$ (Proposition 3.9). □

Before leaving the complex case we collect some facts which are essentially already established but which will be useful in what follows.

PROPOSITION 4.14. *Let X be a compact subset of* $\mathbf{C}$ *and R any discrete subring of* $\mathbf{C}$. *Then* (i) *through* (iv) *of the following are equivalent if X is infinite and R has rank* 2. *In general,* (i) *is equivalent to* (ii), (ii) *implies* (iii), (iii) *is equivalent to* (iv), *and* (iv) *implies* (v):

(i) $d(X) \geqslant 1$;

(ii) *If* $p \in \mathbf{C}[z]$ *and p is monic then* $\|p\|_X \geqslant 1$;

(iii) $B(X, R) = \{0\}$;

(iv) $J(X, R) = X$;

(v) $R[z]$ *is uniformly closed in* $C(X)$.

PROOF. The equivalence of (i) and (ii) is Theorem 2.11(i). To establish that (ii) implies (iii) we prove the contrapositive. Let $q \in B(X, R)$ with $q \not\equiv 0$ on X. Then q has a nonzero leading coefficient a. Since R is discrete we have, by a remark following Definition 1.1, $|a| \geqslant 1$. Thus the polynomial $a^{-1}q$ is in $\mathbf{C}[z]$, is monic, and $\|a^{-1}q\|_X < 1$. The equivalence of (iii) and (iv) is obvious from the definition of $J(X, R)$. We see that (iv) implies (v) immediately from Proposition 3.7.

It remains to assume that X is infinite and establish that (iii) implies (ii). We do so by proving the contrapositive. Thus let p be monic and $\|p\|_X < 1$. By Theorem 3.8 there exists q in $R[z]$ with $q \neq 0$ and $\|q\| < 1$. Thus $q \in B(X, R)$. Since q is not the zero polynomial, it has at most finitely many zeros. Thus it is not identically zero on X, so $B(X, R) \neq \{0\}$. $\square$

CHAPTER 5

REAL CASE

In this chapter we solve the problem of approximation by integral polynomials where X is any compact subset of the real line and the coefficients are the rational integers $\mathbf{Z}$ or any nonzero discrete subring R of the reals $\mathbf{R}$. The problem is solved in the sense that we characterize the functions which can be so approximated, much as in the preceding chapter. The results follow directly from those of the preceding chapter.

The results in the case where the ring of coefficients is any nonzero discrete subring R of $\mathbf{R}$ can be deduced from the case where the ring of coefficients is $\mathbf{Z}$. This is essentially because such a ring has the form $n\mathbf{Z}$ for some positive integer n, as follows. Let $n = \min\{r \in R: r > 0\}$. Then certainly $n\mathbf{Z} \subset R$ and to show equality suppose $r' \in R \setminus n\mathbf{Z}$. Then there exists an integer m such that $nm < r' < n(m+1)$; hence $0 < r' - nm < n$ which contradicts the definition of n. Thus $n\mathbf{Z} = R$ and to see that n is an integer, notice that we must have $n^2 = nm$ for some integer m; hence $n = m$.

We first show that the two different definitions for the algebraic kernel of a compact subset of the reals are, in a sense, unnecessary.

PROPOSITION 5.1. *Let X be a compact subset of $\mathbf{R}$. For any imaginary quadratic field L, we have*

$$J(X, \mathbf{Z}) = J(X, I_L).$$

PROOF. Since $B(X, \mathbf{Z}) \subset B(X, I_L)$, the inclusion $J(X, I_L) \subset J(X, \mathbf{Z})$ is clear from the definition. On the other hand, let $x_0 \in J(X, \mathbf{Z})$ and $p \in B(X, I_L)$. Then $\|p\| < 1$, so for some positive integer n, $\|p^n\| = \|p\|^n < \frac{1}{2}$. Then we have $\|\mathrm{Re}(p^n)\| < \frac{1}{2}$ and $\|\mathrm{Im}(p^n)\| < \frac{1}{2}$. Furthermore, from Proposition 1.6 we see that $2\,\mathrm{Re}(p^n)$ and $(2/\sqrt{|d|}\,)\mathrm{Im}(p^n)$ are in $\mathbf{Z}[x]$, where $L = \mathbf{Q}(\sqrt{d}\,)$ with d a square-free integer and $\mathrm{Re}(p^n)$ (resp. $\mathrm{Im}(p^n)$) denotes the polynomial obtained by replacing the coefficients of p^n by their real (resp. imaginary) parts. Also

$$\|2\,\mathrm{Re}(p^n)\| = 2\|\mathrm{Re}(p^n)\| < 1$$

and

$$\left\|(2/\sqrt{|d|}\,)\mathrm{Im}(p^n)\right\| < (1/|d|)^{1/2} \leqslant 1$$

and so by definition of $J(X, \mathbf{Z})$, $2\,\mathrm{Re}(p'')(x_0) = 0$ and $(2/\sqrt{|d|}\,)\mathrm{Im}(p'')(x_0) = 0$. But $p''(x_0) = (\mathrm{Re}(p''))(x_0) + (\mathrm{Im}(p''))(x_0)$ and so $p''(x_0) = 0$, which implies that $p(x_0) = 0$. Hence $x_0 \in J(X, I_L)$ and $J(X, \mathbf{Z}) \subset J(X, I_L)$. □

Before proving the next proposition we note that if X is a compact subset of R, then it is Lavrent'ev by Proposition 4.2 or by Weierstrass' theorem.

PROPOSITION 5.2. *Let X be a compact subset of $\mathbf{R}$ with $d(X) < 1$. Then for any imaginary quadratic field L,*

$$J_0(X, \mathbf{Z}) = J_0(X, I_L).$$

PROOF. If $x_0 \in J_0(X, \mathbf{Z})$, then x_0 is a root of a monic polynomial $p \in \mathbf{Z}[x]$ which has all of its roots in X. Thus z_0 is integral over I_L. The minimal polynomial q of x_0 over L is then an element of $I_L[z]$, monic, irreducible, and divides p so that the roots of q all lie in X. Thus $x_0 \in J_0(X, I_L)$.

For the reverse inclusion notice that $J_0(X, I_L) = J(X, I_L) = J(X, \mathbf{Z})$ by Theorem 4.10 and Proposition 5.1. This shows, in particular, that $J_0(X, I_L)$ is independent of the choice of L. Suppose that $L = \mathbf{Q}(i)$ where $i^2 = -1$. Then I_L is the set of Gaussian integers. If $x_0 \in J_0(X, I_L)$ then it is a root of a monic, irreducible p in $I_L[z]$, having all of its roots in X. The coefficients of p, being simply the elementary symmetric polynomials in the roots, are in $I_L \cap \mathbf{R}$. But $I_L \cap \mathbf{R} = \mathbf{Z}$, so $x_0 \in J_0(X, \mathbf{Z})$. Since x_0 was any element of $J_0(X, I_L)$, $J_0(X, I_L) \subset J_0(X, \mathbf{Z})$. □

From Propositions 5.1 and 5.2 and Theorem 4.9, we have the following result.

THEOREM 5.3. *If X is any compact subset of the real line $\mathbf{R}$ with $d(X) < 1$, then $J(X, \mathbf{Z})$ is finite and*

$$J(X, \mathbf{Z}) = J_0(X, \mathbf{Z}).$$

A natural question at this point is whether or not the hypothesis $d(X) < 1$ can be dropped from Theorems 5.3 or 4.10. We see that it cannot be dropped in either case by the following argument.

Let L be an imaginary quadratic field and X any uncountable compact subset of C with $d(X) \geq 1$. We know that $B(X, I_L) = \{0\}$ by the comments following Definition 3.2; so $B(X, \mathbf{Z}) = \{0\}$. This implies that $J(X, I_L) = J(X, \mathbf{Z}) = X$ by definition. On the other hand, every element of $J_0(X, \mathbf{Z})$ (resp. $J_0(X, I_L)$) is algebraic over $\mathbf{Q}$ and so $J_0(X, \mathbf{Z})$ (resp. $J_0(X, I_L)$) is countable, hence not equal to $X = J(X, \mathbf{Z}) = J(X, I_L)$.

Another question is whether or not it is necessary to allow polynomials with complex coefficients when seeking the Čebyšev polynomials $t_n(x, X)$ for $X \subset \mathbf{R}$. This is not necessary since the Čebyšev polynomials as defined have real coefficients in this case as follows. Since x is real on X we have for each $x \in X$

$$|t_n(x)| \geq |\mathrm{Re}(t_n(x))| = |(\mathrm{Re}\, t_n)(x)|;$$

hence $\|t_n\|_X \geq \|\mathrm{Re}\, t_n\|_X$. Since t_n is monic and of degree n, $\mathrm{Re}\, t_n$ is also. Thus by Proposition 2.3, $t_n = \mathrm{Re}\, t_n$ as functions on X. If X has n or more elements $t_n = \mathrm{Re}\, t_n$ as polynomials. If not, then by the Definition 2.4 we see that t_n has real coefficients. As a consequence of this argument we see that if we define, for $X \subset \mathbf{R}$ and $\operatorname{card} X \geqslant n$, $t_n(x, X)$ to be the monic polynomial with real coefficients with least supremum norm on X, then this is compatible with Definition 2.4. The case $\operatorname{card} X < n$ is handled as before. If we now define $d(X)$ from the uniform norms of the t_n as before then we will have a definition which nowhere involves complex numbers.

The main result of this chapter is the following.

THEOREM 5.4. *If X is a compact subset of $\mathbf{R}$ then a function f in $C(X, \mathbf{R})$ is $\mathbf{Z}$-approximable if and only if f is $\mathbf{Z}$-interpolable on $J(X, \mathbf{Z})$. If $d(X) \geqslant 1$ then f in $C(X, \mathbf{R})$ is $\mathbf{Z}$-approximable on X if and only if f is already an element of $\mathbf{Z}[x]$.*

PROOF. If f is $\mathbf{Z}$-approximable then it is $\mathbf{Z}$-interpolable on $J(X, \mathbf{Z})$ by Proposition 3.7. Conversely, first suppose $d(X) < 1$. Assume that f is in $C(X, \mathbf{R})$ and that f is interpolated on $J(X, \mathbf{Z})$ by $p \in \mathbf{Z}[x]$. Since it suffices to approximate $f - p$ we can assume without loss of generality that $f \equiv 0$ on $J(X, \mathbf{Z})$. Let L be any imaginary quadratic field (the Gaussian numbers $\mathbf{Q}(i)$ for example). Then by Proposition 5.1, $J(X, I_L) = J(X, \mathbf{Z})$ and since $f \equiv 0$ on $J(X, \mathbf{Z})$, f is I_L-interpolable on $J(X, I_L)$. Thus for $\varepsilon > 0$ there exists $p \in I_L[z]$ such that

$$\|p - f\|_X < \varepsilon/2 \tag{$*$}$$

by Theorems 4.10 and 4.7. Then for any $x \in X$

$$\varepsilon/2 > |\operatorname{Im}(p(x) - f(x))| = |\operatorname{Im} p(x)| = |(\operatorname{Im} p)(x)|$$

where $\operatorname{Im} p$ (resp. $\operatorname{Re} p$) is obtained from p by replacing each coefficient by its imaginary (resp. real) part. Since $p = \operatorname{Re} p + i \operatorname{Im} p$, $\|p - \operatorname{Re} p\|_X < \varepsilon/2$, and by $(*)$ we have

$$\|\operatorname{Re} p - f\|_X < \varepsilon.$$

If $d(X) \geqslant 1$ then the second statement of the theorem is a special case of Theorem 2.11(ii). Also, by Proposition 4.14, if $d(X) \geqslant 1$, then $J(X, \mathbf{Z}) = X$ and the other direction of the first statement now follows in this case. □

REMARK 5.5. We note here that the first statement of Theorem 5.4 is a special case of the very general Theorem 9.2. There is still a third method of proving Theorem 5.4 which is the most direct of all. This method is an extension of the method of proof of Pál's result in our introduction and was communicated to us privately by David Cantor. We relate it now.

If f is $\mathbf{Z}$-approximable on X, then it is $\mathbf{Z}$-interpolable on $J(X)$ by Proposition 3.7 as before. Conversely, let $f \in C(X, \mathbf{R})$ and interpolable by q_0 on $J(X)$. Then $f - q_0 \equiv 0$ on $J(X)$ and it suffices to approximate $f - q_0$; hence we assume without loss of generality that $f \equiv 0$ on $J(X)$. If $J(X) = X$ we are obviously done; so, assume not. Then there exists $\tilde{q}$ in $B(X, \mathbf{Z})$ with $\tilde{q} \not\equiv 0$ on X. Let $Z_{\tilde{q}} \setminus J(X) = \{x_1, \ldots, x_m\}$ and for each x_i pick q_{x_i} in $B(X, \mathbf{Z})$ with $q_{x_i}(x_i) \neq 0$.

For any even integer n we can define q in $Z[x]$ by $q = \tilde{q}^n + q_{x_1}^n + \cdots + q_{x_m}^n$ and have

$$J(X) = Z_q \cap X. \tag{1}$$

For large enough n we also have

$$\delta = \max\{\|q(x)\|_X, \|xq(x)\|_X\} < 1$$

where the equality serves to define δ. Let ε be any positive number. The series $\Sigma_{i+j>0}\delta^{i+j}$ is convergent (in fact it sums to $(1-\delta)^{-2}$); hence there exists a positive integer k such that $\Sigma_{i+j\geq k}\delta^{i+j} < \varepsilon/2$. Let X_0 be the compact Hausdorff space obtained from X by identifying the points of $J(X)$. Then we can consider f and q to be defined on X_0. By (1), if y_1 and y_2 are distinct points in X_0, then not both $q(y_1)$ and $q(y_2)$ are zero. Thus either $q^k(y_1) \neq q^k(y_2)$ or $y_1q^k(y_1) \neq y_2q^k(y_2)$; that is, the pair $\{q^k(x), xq^k(x)\}$ forms a point separating subset of $C(X_0, \mathbf{R})$ and, by a form of the Stone-Weierstrass theorem (Naimark [**64**, Theorem 1, p. 32]), there is a polynomial p in two variables with real coefficients and no constant term such that for every x in X_0, hence $x \in X$,

$$|f(x) - p(q^k(x), xq^k(x))| < \varepsilon/2. \tag{2}$$

If we let $[p]$ denote the polynomial p with each coefficient replaced by its integral part, then for every x in X,

$$\|p(q^k(x), xq^k(x)) - [p](q^k(x), xq^k(x))\| \leq \sum_{i+j\geq k} \|q^k(x)\|^i \|xq^k(x)\|^j$$

$$\leq \sum_{i+j\geq k} \delta^{i+j} < \varepsilon/2. \tag{3}$$

By (2) and (3),

$$\|f(x) - [p](q^k(x), xq^k(x))\| < \varepsilon$$

and $[p](q^k(x), xq^k(x)) \in \mathbf{Z}[x]$. □

Since the transfinite diameter of a line segment is one-fourth its length (Example 2.15), we have the following.

COROLLARY 5.6. *If $X = [a, b]$ and $b - a \geq 4$ then f is $\mathbf{Z}$-approximable on X if and only if f is already an element of $\mathbf{Z}[x]$.*

Using essentially the same arguments as for their respective counterparts in the complex case, Theorems 4.8 and 4.13, the following can be proved.

PROPOSITION 5.7. *Let X be a compact subset of $\mathbf{R}$ with $d(X) < 1$. Then an element f of $C(X, \mathbf{R})$ is $\mathbf{Z}$-approximable on X if and only if the Lagrange interpolating polynomial for f on $J(X, \mathbf{Z})$ is an element of $\mathbf{Z}[x]$.*

PROPOSITION 5.8. *Let X be a compact subset of $\mathbf{R}$ with $d(X) < 1$. Then $\mathbf{Z}[x]$ is uniformly closed in $C(X, \mathbf{R})$ if and only if $J(X, \mathbf{Z}) = X$. In particular, if X is infinite, then $\mathbf{Z}[x]$ is not uniformly closed in $C(X, \mathbf{R})$.*

PROPOSITION 5.9. *Let X be a compact subset of* **R** *with $d(X) < 1$. Then $J_0(X, \mathbf{Z}) = J(X, \mathbf{Z})$ is finite and there exists q in $B(X, \mathbf{Z})$ such that $J_0(X, \mathbf{Z}) = Z_q \cap X$.*

PROOF. Let A be the Gaussian integers $\mathbf{Z} + i\mathbf{Z}$ and apply Theorem 4.9 to get q in $B(X, A)$ with $J_0(X, A) = Z_q \cap X$. Let Re q and Im q be the polynomials defined as in the proof of Proposition 5.1. Then the polynomial $\tilde{q} = (\mathrm{Re}\, q)^2 + (\mathrm{Im}\, q)^2$ is in $B(X, \mathbf{Z})$ and $J_0(X, A) = Z_{\tilde{q}} \cap X$. By Proposition 5.2, $J_0(X, A) = J_0(X, \mathbf{Z})$. Also, $J_0(X, \mathbf{Z}) = J(X, \mathbf{Z})$ by Theorem 5.3. □

In the case of L_p approximation, we have the following result. It is interesting to note that there are no longer any arithmetic conditions for approximability, but there is a condition on the size of the domain.

THEOREM 5.10. *Let $1 \leqslant p < \infty$. If $b - a < 4$, then every f in $L_p[a, b]$ can be approximated in L_p norm by elements of $Z[x]$. If $b - a \geqslant 4$, then no element of $L_p[a, b] \setminus Z[x]$ can be so approximated.*

PROOF. The last statement is Theorem 2.12. If $b - a < 4$, then we know from Proposition 5.9 and Example 2.15 that $J[a, b]$ is finite, hence λ-null, where λ is Lebesgue measure. The result will follow from Theorem 9.11. □

REMARK 5.11. In closing this chapter we mention a slight generalization of the above results which can be proved by the foregoing techniques, but which follows immediately from the adelic case to be treated next. We need not restrict attention to compact subsets X of **R**, but could handle any compact subset of **C** with void interior, connected complement, and satisfying $\overline{X} = X$. In place of $C(X, \mathbf{R})$ we can take $C_0(X) = \{f \in C(X) | f(\bar{x}) = \overline{f(x)}, x \in X\}$. See Remark 6.5 in the next chapter.

CHAPTER 6

ADELIC CASE

In this case we start with an algebraic number field K and consider approximation in several uniform norms simultaneously by the class of integral polynomials $K^T[x]$ defined as follows.

Let Ω be the set of all equivalence classes of valuations on K. (Two valuations are equivalent if they give rise to the same metric topology on K.) If $v \in \Omega$ then every element of v gives rise to the same metric completion of K which we denote by K_v. Denote the Archimedean elements of Ω by Ω_∞ and take T to be any finite set satisfying $\Omega_\infty \subseteq T \subset \Omega$. From each equivalence class v in Ω we pick a canonical valuation $|\cdot|_v$ as follows. If an element of v, when restricted to $\mathbf{Q}$, is the usual absolute value $|\cdot|$ then $K_v = \mathbf{R}$ or $\mathbf{C}$ and we take $|\cdot|_v = |\cdot|$ or $|\cdot|^2$, respectively. If not, then the restriction corresponds to a p-adic valuation with associated prime $\mathfrak{p}$ and we choose $|\cdot|_v$ to be that element of v with $|\mathfrak{p}|_v = \mathfrak{p}^{-N_v}$ where N_v is the degree of the field extension $K_v/\mathbf{Q}_\mathfrak{p}$.

As noted in Chapter 1, the ring of integers in this case is defined to be $K^T = \{k \in K\colon |k|_v \leqslant 1, \text{ all } v \in \Omega \setminus T\}$. The ring K^T is called the *ring of T-integers* of K and the ring of integral polynomials in this case is $K^T[x]$, i.e., the polynomials whose coefficients are T-integers. Properties of the ring K^T are studied in David Cantor [**65**].

For each $v \in T$ let $\tilde{K}_v$ be the algebraic closure of K_v. Since K is dense in K_v the valuation $|\cdot|_v$ on K has a unique extension to K_v and since $\tilde{K}_v$ is the union of the finite extensions between K_v and $\tilde{K}_v$, $|\cdot|_v$ has a unique extension to $\tilde{K}_v$ (O'Meara [**63**, 14 : 1]). Topologize $\tilde{K}_v$ by the resulting metric and let X_v be a compact subset of $\tilde{K}_v$. We define $C(X_v, \tilde{K}_v)$ to be the space of all continuous $\tilde{K}_v$-valued functions on X_v. It is an algebra over $\tilde{K}_v$ when the algebraic operations are defined pointwise, as usual. We define a norm on $C(X_v, \tilde{K}_v)$ by

$$\|f\|_{X_v} = \sup_{x \in X_v} |f(x)|_v.$$

Notice that $K^T[x] \subset C(X_v, \tilde{K}_v)$ in the obvious sense. Suppose $f_v \in C(X_v, \tilde{K}_v)$ for each v in T. We say that the family $\{f_v\}_{v\in T}$ is *approximable* (*T-approximable*) if given any $\varepsilon > 0$ there exists q in $K^T[x]$ such that

$$\|q - f_v\|_{X_v} < \varepsilon \quad \text{all } v \in T.$$

The main purpose of this chapter is to give conditions on the family $\{f_v\}$ in order that it be approximable.

We first show that we can assume a certain symmetry for the sets X_v without loss of generality. Let $\mathrm{Aut}(\tilde{K}_v/K_v)$ denote the automorphisms of the field extension $\tilde{K}_v/K_v$. For any $\sigma \in \mathrm{Aut}(\tilde{K}_v/K_v)$ it is clear that every $q \in K^T[x]$ is invariant under σ $(\sigma q(x) = q(\sigma x))$; hence, if $\{f_v\}_{v\in T}$ is approximable and $\sigma x \in X_v$, then $\sigma f_v(x) = f_v(\sigma x)$. By David Cantor [**67**, Theorem 7], if v is non-Archimedean and $\{f_v\}_{v\in T}$ is approximable, then each f_v has a continuous extension f_v' to the compact set $X_v' = [\mathrm{Aut}(\tilde{K}_v/K_v)]X_v$ satisfying $\sigma f_v'(x) = f_v'(\sigma x)$, all x in X_v' and all σ in $\mathrm{Aut}(\tilde{K}_v/K_v)$. The same holds in the Archimedean case as follows. If v is Archimedean either $\mathrm{Aut}(\tilde{K}_v/K_v)$ is the trivial group and we are done or it consists of two elements: the identity and complex conjugation. In the latter case let $\tilde{f}_v$ be defined on $\overline{X}_v$ by

$$\tilde{f}_v(x) = \bar{f}_v(\bar{x}).$$

It is clear that $\tilde{f}_v$ is a continuous function on $\overline{X}_v$. We define the extension f_v' of f_v to $\overline{X}_v \cup X_v = [\mathrm{Aut}(\tilde{K}_v/K_v)]X_v$ by

$$f_v'(x) = \begin{cases} f_v(x) & \text{if } x \in X_v, \\ \tilde{f}_v(x) & \text{if } x \in \overline{X}_v. \end{cases}$$

Then f_v' is well defined since f_v and $\tilde{f}_v$ agree on $\overline{X}_v \cap X_v$ as is easily checked. Since f_v and $\tilde{f}_v$ are continuous on the closed sets X_v and $\overline{X}_v$, respectively, the extension f_v' is continuous. The property $f_v'(\bar{x}) = \bar{f}_v'(x)$ is also easily checked. Thus we will assume from this point on that each X_v is *invariant*, i.e., that $\mathrm{Aut}(\tilde{K}_v/K_v)X_v = X_v$.

For such an X_v we denote by $C_0(X_v)$ the set of elements of $C(X_v, \tilde{K}_v)$ which are left fixed by the action of $\mathrm{Aut}(\tilde{K}_v/K_v)$. Thus f in $C(X_v, \tilde{K}_v)$ is an element of $C_0(X_v)$ if and only if $f(\sigma x) = \sigma f(x)$, $\sigma \in \mathrm{Aut}(\tilde{K}_v/K_v)$, for every x in X_v.

A compact invariant subset X_v of $\tilde{K}_v$ is said to be a *Lavrent'ev set* (in D. Cantor [**69**]: "approximation set") if $K_v[x]$ is dense in $C_0(X_v)$ in the uniform norm $\|\cdot\|_{X_v}$. If v is non-Archimedean then every compact invariant subset of $\tilde{K}_v$ is an approximation set by D. Cantor [**67**, Theorem 8]. If v is Archimedean then $\tilde{K}_v = \mathbf{C}$ and it is essentially the classical result due to Lavrent'ev [**36**] that a compact invariant subset X_v of $\tilde{K}_v$ is Lavrent'ev if and only if it has void interior and connected complement.

The concept of the transfinite diameter $d(X_v)$ for compact invariant subsets X_v of $\tilde{K}_v$ has been given in Definition 2.17. The sets J^0 and J_v^0 were defined in Definition 3.11.

If $f_v \in C_0(X_v)$ for each v in T, then we say that the family $\{f_v\}_{v\in T}$ is *interpolable* if there exists q in $K^T[x]$ such that

$$f_v(x) = q(x) \quad \text{all } x \in J_v^0, \quad v \in T.$$

We can now state the main theorem of this chapter.

THEOREM 6.1 *For each v in T let X_v be a Lavrent'ev subset of $\tilde{K}_v$ and f_v an element of $C_0(X_v)$. Then*

(i) *if $\Pi_{v\in T}\delta(X_v) < 1$, the family $\{f_v\}_{v\in T}$ is approximable if and only if it is interpolable;*

(ii) *if $\Pi_{v\in T}\delta(X_v) \geqslant 1$, the family $\{f_v\}_{v\in T}$ is approximable if and only if there exists q in $K^T[x]$ such that*

$$f_v(x) = q(x), \qquad x \in X_v, v \in T.$$

PROOF. We will prove (ii) and that interpolability is necessary for approximability in (i). For the converse of the latter we refer the reader to D. Cantor [**69**] where it is established after a number of lemmas.

In (ii) suppose that $\{f_v\}_{v\in T}$ is approximable. Then there exist q_1, q_2 in $K^T[x]$ such that $\|q_i - f_v\|_{X_v} < \frac{1}{4}$, $v \in T$, $i = 1, 2$. We claim that q_1 and q_2 are identical functions on each X_v. Suppose not. Then setting $q = q_1 - q_2$ we have

$$\|q\|_{X_v} < 1, \qquad v \in T, \tag{1}$$

and q is not a constant polynomial since if it were this constant c would satisfy $|c|_v \leqslant 1$, $v \in \Omega \setminus T$, since it is a T-integer and $|c|_v < 1$, $v \in T$, by (1); hence $c = 0$ by the product formula (cf. O'Meara [**63**, 33:1]). Let d be the degree of q and a its leading coefficient. Setting $\tilde{q} = a^{-1}q$ we have that $\tilde{q}$ is monic and $\|\tilde{q}\|_{X_v} < |a^{-1}|_v$, $v \in T$. As is easily seen from Definition 2.17, this implies that

$$d(X_v) < |a^{-1}|_v, \qquad v \in T. \tag{2}$$

Since a is a nonzero T-integer, $\Pi_{v\in T}|a|_v \geqslant 1$. Thus by (2), $\Pi_{v\in T}d(X_v) < 1$ which is a contradiction. Thus $q_1 \equiv q_2$ on X_v, $v \in T$, whenever $q_1, q_2 \in K^T[x]$ and $\|q_i - f_v\|_{X_v} < \frac{1}{4}$, $i = 1, 2$. It is clear from this that we can take for q in (ii) any element of $K^T[x]$ satisfying $\|q - f\|_{X_v} < \frac{1}{4}$, all v in T. The converse of (ii) is obvious.

Suppose we are in the situation of (i) and $\{f_v\}_{v\in T}$ is approximable. We proceed as above and obtain q satisfying (1). We will be done once we show that (1) implies $q \equiv 0$ on J_v^0 all $v \in T$ and this is clear from the following.

PROPOSITION 6.2. *Let $q \in K^T[x]$ and $\|q\|_{J_v^0} < 1$, all $v \in T$. Then $q \equiv 0$ on J^0.*

PROOF. Let $\alpha \in J^0$, $L = K(\alpha)$, $N_{L/K}$ be the norm from L to K and $S_v = \mathrm{Emb}_K(L, \tilde{K}_v)$. Then

$$\prod_{v\in T} |N_{L/K}(q(\alpha))|_v = \prod_{v\in T}\prod_{\sigma\in S_v} |\sigma(q(\alpha))|_v = \prod_{v\in T}\prod_{\sigma\in S_v} |q(\sigma(\alpha))|_v < 1 \quad (*)$$

since $\|q\|_{X_v} < 1$, each v in T, and each σ in S_v can be extended to $\tilde{K} = \tilde{L}$ by field theory; hence $\sigma(\alpha) \in X_v$, for all $\sigma \in S_v$, $v \in T$. Also, since α is integral over K^T and $q \in K^T[x]$, $q(\alpha)$ is integral over K^T. Thus $N_{L/K}(q(\alpha))$ is integral over K^T and there is a monic $\tilde{q}$ in $K^T[x]$ with $\tilde{q}(q(\alpha)) = 0$. As is well known, the norm $N_{L/K}q((\alpha))$ is (to within sign) one of the coefficients of the minimal polynomial p of $q(\alpha)$ over K. Since $\tilde{q}(q(\alpha)) = 0$ we have that p divides $\tilde{q}$. Applying O'Meara [**63**, 13:7] at each $v \in \Omega \setminus T$ we see that the coefficients of p

are all T-integers; hence $N_{L/K}(q(\alpha))$ is a T-integer. By the product formula and (*) we see that $N_{L/K}(q(\alpha)) = 0$; hence $q(\alpha) = 0$. $\square$

The following analogues of our Theorems 4.9 and 4.10 hold and are proved similarly (D. Cantor [**69**, p. 10]).

THEOREM 6.3. *For each $v \in T$ let X_v be a Lavrent'ev subset of $\tilde{K}_v$ and $\Pi_{v \in T}\delta(X_v) < 1$. Then J^0 is finite and there exists q in $K^T[x]$ with $J_v^0 = X_v \cap Z_{v,q}$ and $\|q\|_{X_v} < 1$, for all $v \in T$. Here $Z_{v,q}$ denotes the roots of q in K_v.*

THEOREM 6.4. *Under the hypotheses of Theorem 6.3, $J^0 = \bigcap_{q \in S} Z_q$ where S is the set of all q in $K^T[x]$ with $\|q\|_{X_v} < 1$ for all v in T and Z_q denotes the roots of q in K.*

In the remainder of this chapter we will show how the results in the adelic case specialize to those in the real and complex cases by a proper choice of K and T.

In order to recover the complex case one may proceed as follows. Let $K = L$, an imaginary quadratic field and $T = \Omega_\infty$. Then the set of Archimedean elements Ω_∞ of Ω consists of exactly one element (O'Meara [**63**, 15:8]) which we will denote by ∞. If we consider L as a subfield of $\mathbf{C}$ in the usual way, then our canonical valuation from ∞ is $|\cdot|^2$ where $|\cdot|$ is the ordinary modulus for complex numbers. It is clear that "approximability" with respect to this norm is equivalent to "approximability" with respect to $|\cdot|$. The square on the modulus is a result of our normalization which is necessary to validate the product formula. It is easy to see that L is dense in $\mathbf{C}$ under the topology induced by $|\cdot|^2$ and since $\mathbf{C}$ is algebraicly closed $L_\infty = \mathbf{C} = \tilde{L}_\infty$. Thus $\mathrm{Aut}(\tilde{L}_\infty/L_\infty)$ is the trivial group here and every compact subset X of $L_\infty = \mathbf{C}$ is invariant. We have already noted in Example 1.15 that in this case ($T = \Omega_\infty$) the ring of T-integers K^T is identical with I_L. If $\alpha \in \tilde{L}$ then we know from field theory that $(\mathrm{Emb}_L(\tilde{L}, \mathbf{C}))(\alpha)$ is the set of conjugates of a over L. Thus if X_∞ is a compact subset of $\mathbf{C}$, the set J^0 in Definition 3.11 is identical to the set $J_0(X_\infty)$ as defined in Definition 3.4. In view of the above correspondences we see that, taking $A = I_L$ in Chapter 4, the Theorems 4.7, 4.9, and 4.10 follow from the present Theorems 6.1, 6.3, and 6.4, respectively. To obtain these theorems of Chapter 4 in full generality we must also consider the case when A is a discrete subring of $\mathbf{C}$ with rank 2 and not equal to I_L for some imaginary quadratic field L but this generalization is easily made in view of Propositions 1.10 and 1.11.

REMARK 6.5. We have already noted how the real case follows from the complex case. It is also possible to obtain a slight generalization of it from the adelic case as follows. Let $K = \mathbf{Q}$, the rational numbers, and $T = \Omega_\infty$. Then again, Ω_∞ is a singleton, namely $\{\infty\}$, where the canonical element of ∞ is the usual absolute value $|\cdot|$ on the real numbers restricted to $\mathbf{Q}$ (O'Meara [**63**, 12:1]). It is clear that $\mathbf{Q}_\infty = \mathbf{R}$ and $\tilde{\mathbf{Q}} = \mathbf{C}$. Thus $\mathrm{Aut}(\tilde{K}_\infty/K_\infty) = \{1, \sigma\}$ where σ is ordinary complex conjugation and 1 the identity map. Thus a compact subset X of $\tilde{K}_\infty = \mathbf{C}$ is invariant if and only if it is symmetric about the real axis, and this certainly includes the case of a compact subset of the real numbers. Also, the elements of $C_0(X_\infty)$ must be real valued if $X_\infty \subset \mathbf{R}$ since then $f_\infty(x) = f_\infty(\bar{x})$ and $f_\infty \in C_0(X_\infty)$ implies $f_\infty(\bar{x}) = \overline{f_\infty(x)}$. The ring K^T is simply $\mathbf{Z}$ in this situation as was pointed out in Example 1.15. If α is algebraic over $\mathbf{Q}$, then we know from field theory that $(\mathrm{Emb}_{\mathbf{Q}}\{\tilde{\mathbf{Q}}, \mathbf{C}\})(\alpha)$ is the set of conjugates of α over $\mathbf{Q}$. Thus for any compact subset X_∞ of $\mathbf{C}$, the set J^0 in Definition 3.11 is identical to the set $J_0(X_\infty)$ as defined in Definition 3.4. In view of these correspondences, Theorems 5.4 and 5.3 follow from Theorems 6.1 and 6.4, respectively.

In deducing the real case from the adelic case we actually obtain a slight generalization as follows. The set X_∞ need not be contained in $\mathbf{R}$ but merely $X_\infty = \bar{X}_\infty$. Likewise the functions in $C_0(X_\infty)$ are not necessarily real valued if $X_\infty \not\subset \mathbf{R}$, but we must have $f(\bar{x}) = \overline{f(x)}$ for each $f \in C_0(X_\infty)$ and $x \in X_\infty$. It is easy to see that such an X_∞ is Lavrent′ev (i.e., $\mathbf{C}[x]$ is dense in $C_0(X_\infty)$) if and only if it has void interior and connected complement. This follows from the classical result of Lavrent′ev **[36]** which says that the polynomials with complex coefficients are uniformly dense in $\mathbf{C}(X)$ whenever X is a compact subset of the complex plane $\mathbf{C}$ with void interior and connected complement.

CHAPTER 7

COMPLEX CASE II: NONVOID INTERIOR

In Chapter 4 we characterized the approximable functions on a compact subset X of $\mathbf{C}$ in the case where $d(X) \geqslant 1$ or $d(X) < 1$ and X has void interior and connected complement. In the present chapter we address the same problem on sets X with interior and $d(X) < 1$. Some of the methods work when X is a compact subset of complex n-space ($\mathbf{C}^n$), and we present them at this level of generality.

We first establish some additional notation. Throughout this chapter X will be a compact subset of $\mathbf{C}^n$, $n \geqslant 1$. The symbol z will stand for an n-tuple of complex numbers $(z_1, \ldots, z_n)$ in $\mathbf{C}^n$. If R is any subring of $\mathbf{C}$, $R[z]$ will denote $R[z_1, \ldots, z_n]$, the ring of polynomials (restrictions to X of polynomial functions, to be precise) in $z_1, \ldots, z_n$ with coefficients in R. When z is an element of $\mathbf{C}^n$ we define $|z|$ by $|z| = (|z_1|^2 + \cdots + |z_n|^2)^{1/2} \geqslant 0$. We write $\Sigma_{k \in \mathbf{N}^n} a_k z^k$ to denote

$$\sum_{k_1, \ldots, k_n = 0}^{\infty} a_{k_1, \ldots, k_n} z_1^{k_1} \cdots z_n^{k_n}.$$

If f is a complex valued function on X and R a subring of $\mathbf{C}$, then we say that f is R-approximable on X if for any $\varepsilon > 0$ there exists $p \in R[z]$ such that

$$\|f - p\|_X = \sup_{z \in X} |f(z) - p(z)| < \varepsilon.$$

The notion of polynomial convexity is important in any treatment of uniform approximation by polynomials on compact subsets of $\mathbf{C}^n$.

DEFINITION 7.1. Let X be a compact subset of $\mathbf{C}^n$. The *polynomial hull* $h(X)$ of X is defined by

$$h(X) = \{z \in \mathbf{C}^n : |p(z)| \leqslant \|p\|_X, \text{all } p \in \mathbf{C}[z]\}.$$

If $X = h(X)$ then X is said to be *polynomially convex*.

By considering the polynomials $p_i(z) = z_i$, $1 \leqslant i \leqslant n$, we see that $|z_i| \leqslant \sup_{z \in X}|z|$, z in $h(X)$, so that $h(X)$ is bounded. It is also an intersection of closed sets, hence closed. Being closed and bounded in $\mathbf{C}^n$, $h(X)$ is compact. Clearly for any p in $\mathbf{C}[z]$ we have $\|p\|_X = \|p\|_{h(X)}$. From this we easily see that $h(h(X)) = h(X)$.

The relationship between ordinary convexity and polynomial convexity is given in the following example which shows that convexity implies polynomial convexity. The converse is false as we will see from Proposition 7.4.

EXAMPLE 7.2. If X is a compact convex subset of $\mathbf{C}^n$, then X is polynomially convex.

PROOF. Let z' be any element of $\mathbf{C}^n \setminus X$. Since $X \subset h(X)$ it suffices to show that there exists a polynomial p' in $\mathbf{C}[z]$ such that $\|p'\|_X < |p'(z')|$. Since X is compact and convex, there exist constants c and ε, $\varepsilon > 0$, and a linear functional f on $\mathbf{C}^n$ such that

$$\sup_{z \in X} \operatorname{Re} f(z) \leqslant c - \varepsilon < c \leqslant \operatorname{Re} f(z'), \tag{1}$$

by the Hahn-Banach theorem. It is well known that f, being a linear functional on $\mathbf{C}^n$, has the form

$$f(z) = \lambda_1 z_1 + \cdots + \lambda_n z_n \tag{2}$$

where $\lambda_i \in \mathbf{C}$ $(1 \leqslant i \leqslant n)$. Thus f is continuous and $f(X \cup \{z'\})$ is compact. Since the MacLauren series for the complex exponential function is uniformly convergent on compact sets, there exists a partial sum $p(w) = 1 + w + \cdots + w^k/k!$ such that

$$\|p(w) - \exp(w)\|_{f(X \cup \{z'\})} < (e^c - e^{c-\varepsilon})/2.$$

Thus

$$\|p(f(z)) - \exp(f(z))\|_{X \cup \{z'\}} < (e^c - e^{c-\varepsilon})/2. \tag{3}$$

From (1) we have $\|\exp(f(z))\|_X \leqslant e^{c-\varepsilon} < e^c < |\exp(f(z'))|$. From this and (3) we obtain

$$\|p(f(z))\|_X < (e^{c-\varepsilon} + e^c)/2 < |p(f(z'))|.$$

Since f has the form given in (2) and p is a polynomial, $p(f(z))$ is a polynomial in $\mathbf{C}[z]$ and we are done. □

In approximation by polynomials we may confine our attention to polynomially convex X as follows.

PROPOSITION 7.3 (WERMER [**61**, p. 73]). *Let X be a compact subset of $\mathbf{C}^n$, $n \geqslant 1$, f a map of X into $\mathbf{C}$, and P any subset of $\mathbf{C}[z]$. If f is uniformly approximable by elements of P, then there exists a unique extension $\tilde{f}$ of f to $h(X)$ such that $\tilde{f}$ is uniformly approximable by elements of P.*

PROOF. Let $\{p_m\}$ be a sequence in P which converges uniformly to f on X. Then by definition of $h(X)$, $\{p_m\}$ is Cauchy in the uniform norm on all of $h(X)$, hence converges uniformly to a limit $\tilde{f}$. It is clear that $\tilde{f}$ is an extension of f to $h(X)$. To see uniqueness, suppose that $\tilde{f}'$ is another extension of f to $h(X)$ and that $\{g_m\}$ is a sequence in P which converges uniformly to $\tilde{f}'$ on $h(X)$. Then for $\varepsilon > 0$ there exists an integer m_ε such that $i, j > m_\varepsilon$ implies $\|p_i - \tilde{f}\|_X = \|p_i - f\|_X < \varepsilon/2$ and $\|g_j - \tilde{f}'\|_X = \|g_j - f\|_X < \varepsilon/2$. It follows that

$$\varepsilon > \|p_i - g_j\|_X = \|p_i - g_j\|_{h(X)};$$

hence both $\{p_m\}$ and $\{g_m\}$ tend to the same limit on $h(X)$. Thus $\tilde{f}' = \tilde{f}$ on $h(X)$ as desired. □

In case $n = 1$ the polynomial hull of X has the following topological characterization.

PROPOSITION 7.4. (HÖRMANDER [**66**, THEOREM 1.3.3]). *Let X be a compact subset of $\mathbf{C}$ and $\{U_i\}_{i=1}^{\infty}$ the set of bounded connected components of $\mathbf{C} \setminus X$. Then we have*

$$h(X) = X \cup \left(\bigcup_{i=1}^{\infty} U_i \right).$$

Thus X is polynomially convex if and only if it has a connected complement.

PROOF. It is clear that $X \subset h(X)$. If U_i is any bounded connected component of $\mathbf{C} \setminus X$, then U_i is closed in the relative topology on $\mathbf{C} \setminus X$. Thus we have $(\mathrm{cl}(U_i) \setminus U_i) \subset X$, and by the maximum modulus principle, if $z_0 \in U_i$ and $p \in \mathbf{C}[z]$ then we have

$$|p(z_0)| \leqslant \sup_{z \in \mathrm{cl}(U_i) \setminus U_i} |p(z)| \leqslant \|p\|_X.$$

Thus $U_i \subset h(X)$ for $i = 1, 2, \ldots$.

Let U be the unbounded component of $\mathbf{C} \setminus X$ and z_0 an element of U. We will be done once we show that z_0 is not in $h(X)$. Since U is open in $\mathbf{C} \setminus X$ and $\mathbf{C} \setminus X$ is open in $\mathbf{C}$, U is open in $\mathbf{C}$. There exists $\varepsilon > 0$ such that $(z_0 + \varepsilon D) \cap X = \varnothing$, where D is the closed unit disk in $\mathbf{C}$. Pick z_1 in $\mathbf{C}$ such that

$$|z_0 - z_1| = \varepsilon/4; \tag{1}$$

that is,

$$1/|z_0 - z_1| = 4/\varepsilon. \tag{2}$$

Since $|z - z_0| > \varepsilon$ for $z \in X$ we have by (1)

$$|z - z_1| > 3\varepsilon/4, \qquad z \in X,$$

and

$$1/|z - z_1| < 4/3\varepsilon, \qquad z \in X. \tag{3}$$

Since $1/(z - z_1)$ is a holomorphic function of z in a neighborhood of $X_0 = X \cup (\bigcup_{i=1}^{\infty} U_i) \cup \{z_0\}$ there exists, by Runge's theorem (Rudin [**74**, 13.7]), a p in $\mathbf{C}[z]$ such that

$$\|p(z) - 1/(z - z_1)\|_{X_0} < 4/3\varepsilon. \tag{4}$$

By (3) and (4), $\|p\|_X < 8/3\varepsilon$, whereas by (2) and (4), $|p(z_0)| > 8/3\varepsilon$. Thus $p(z_0) > \|p\|_X$ which shows that $z_0 \notin h(X)$. □

The Čebyšev polynomials $\{t_m(z, X)\}$ associated with any compact subset X of

C were given in Definition 2.4. They remain essentially the same in passing from X to $h(X)$ as the following result shows.

PROPOSITION 7.5. *Let X be a compact subset of* **C**. *Then $t_m(z, X)$ and $t_m(z, h(X))$ are the same functions on $h(X)$ for all m.*

PROOF. We have

$$\begin{aligned}\|t_m(z, X)\|_{h(X)} &= \|t_m(z, X)\|_X \leqslant \|t_m(z, h(X))\|_X \\ &= \|t_m(z, h(X))\|_{h(X)} \leqslant \|t_m(z, X)\|_{h(X)}. \quad \square\end{aligned}$$

COROLLARY 7.6. *If X is a compact subset of* **C**, *then we have $d(h(X)) = d(X)$, where $d(X)$ is the transfinite diameter (Definition 2.9) of X.*

Before concluding this discussion of polynomial convexity we give two examples which will be needed later.

EXAMPLE 7.7. Let the compact subset X of $\mathbf{C}^2$ be defined by

$$X = \{z \in \mathbf{C}^2: |z_1| = 1 = |z_2|\}.$$

We will show that $h(X) = X_0$ where

$$X_0 = \{z \in \mathbf{C}^2: |z_1| \leqslant 1, |z_2| \leqslant 1\}.$$

By considering the polynomials $p_1(z) = z_1$ and $p_2(z) = z_2$ we see from the definition of $h(X)$ that $h(X) \subset \{z \in \mathbf{C}^2: |z_1| \leqslant 1\}$ and $h(X) \subset \{z \in \mathbf{C}^2: |z_2| \leqslant 1\}$. It remains to show that $X_0 \subset h(X)$. Let z' be an element of C^2 such that $|z_1'| \leqslant 1$ and $|z_2'| \leqslant 1$. Let p be any polynomial in $\mathbf{C}[z]$. Since $z' \in X_0$ we have

$$|p(z')| \leqslant \sup_{z \in X_0} |p(z)| = \sup_{z \in X_0} |p(z_1, z_2)|.$$

But $\sup_{z \in X_0}|p(z_1, z_2)| = \sup_{|z_1| \leqslant 1}(\sup_{|z_2| \leqslant 1}|p(z_1, z_2)|)$ and for each fixed z_1, $p(z_1, z_2)$ is a polynomial in the single complex variable z_2 and so by the maximum modulus principle,

$$\begin{aligned}\sup_{|z_1| \leqslant 1}\Big(\sup_{|z_2| \leqslant 1}|p(z_1, z_2)|\Big) &= \sup_{|z_1| \leqslant 1}\Big(\sup_{|z_2| = 1}|p(z_1, z_2)|\Big) \\ &= \sup_{|z_2| = 1}\Big(\sup_{|z_1| \leqslant 1}|p(z_1, z_2)|\Big) \\ &= \sup_{|z_2| = 1}\Big(\sup_{|z_1| = 1}|p(z_1, z_2)|\Big) = \sup_{z \in X}|p(z)|.\end{aligned}$$

Thus $|p(z')| \leqslant \sup_{z \in X}|p(z)|$ and z' is in $h(X)$ as desired.

It is clear that the idea can be applied to prove that for any positive integer n, if

$$X = \{z \in \mathbf{C}^n: |z_1| = \cdots = |z_n| = 1\},$$

then

$$h(X) = \{z \in \mathbf{C}^n: |z_i| \leqslant 1, 1 \leqslant i \leqslant n\}.$$

Two remarks seem to be in order. First, it is clear that X does not contain the topological boundary of $h(X)$ in these examples (for $n > 1$), although this is always the case for $n = 1$ as we see by Proposition 7.4. Secondly, Proposition 7.4 implies that if $\mathbf{C} \setminus X$ is connected, then $X = h(X)$. This is false for general $X \subset \mathbf{C}^n$ $(n > 1)$ since in the example above $\mathbf{C} \setminus X$ is connected, as is easily seen.

EXAMPLE 7.8. Let the compact subset X of $\mathbf{C}^n$ $(n > 1)$ be defined by $X = \{z \in \mathbf{C}^n: |z_1| = 1, z_2 = \cdots = z_n = 0\}$. Then we have $h(X) = X_0$ where $X_0 = \{z \in \mathbf{C}^n: |z_1| \leqslant 1, z_2 = \cdots = z_n = 0\}$. This is quite close to the previous example and is easily proved by similar techniques.

For the remainder of this chapter, let R be any discrete (Definition 1.1) subring of $\mathbf{C}$. We now give some necessary conditions in order that a complex valued function f on a compact subset X of $\mathbf{C}^n$ be R-approximable on X.

PROPOSITION 7.9. *If f is a complex valued function on a compact subset X of $\mathbf{C}^n$ and f is R-approximable on X then f has a unique extension $\tilde{f}$ to $h(X)$ which is continuous on $h(X)$, holomorphic on $h(X)^\circ$ (where the superscript $^\circ$ denotes "interior"), and R-approximable on $h(X)$.*

The proof is immediate from Proposition 7.3 since uniform limits preserve continuity and holomorphicity.

An example of the use of Proposition 7.9 is the following. Let $X = \{z \in \mathbf{C}: \delta_1 \leqslant |z| \leqslant \delta_2\}$ where $0 < \delta_1 < \delta_2 < 1$ and let the function $z + \frac{1}{2}$ be restricted to X. By Proposition 7.4, $h(X) = \delta_2 D$ where D is the closed unit disk in $\mathbf{C}$. By the principle of holomorphic continuation, the only continuous extension of f to $h(X)$ which is holomorphic on $\delta_2 D^\circ$ is the function $z + \frac{1}{2}$ restricted to $\delta_2 D$. But this function is not R-approximable on $\delta_2 D$ for any discrete ring R since any polynomial $p \in R[z]$ has an element of R as its value at $z = 0$ and every nonzero element of a discrete subring of $\mathbf{C}$ has modulus at least unity. Thus f is not R-approximable on X for any discrete ring R.

In the case $X \subset \mathbf{C}$ it is easy to use Proposition 7.9 as follows. Given a complex valued function f on X we seek an extension f' of f to $h(X)$ which is continuous on $h(X)$ and holomorphic on $h(X)^\circ$. If none such exists, then f is not R-approximable on X. If such an f' exists, then f is R-approximable on X if and only if f' is R-approximable on $h(X)$. This is true since if f is R-approximable on X, then by Proposition 7.9 there exists an R-approximable extension $\tilde{f}$ of f to $h(X)$ which is continuous on $h(X)$ and holomorphic on $h(X)^\circ$. By Proposition 7.4 and the maximum modulus principle $\tilde{f}$ is uniquely determined by f and so $\tilde{f} = f'$ which shows that f' is R-approximable. The converse is obvious.

This procedure is based on the fact that if $X \subset \mathbf{C}$ then there is at most one extension f' of f to $h(X)$ which is continuous on $h(X)$ and holomorphic on $h(X)^\circ$. If $X \subset \mathbf{C}^n$, $n > 1$, the extension need not be unique, as the following example shows.

Let $X = \{z: |z_1| = 1, z_2 = \cdots = z_n = 0\}$. We know from Example 7.8 that $h(X) = \{z: |z_1| \leqslant 1, z_2 = \cdots = z_n = 0\}$. If we define $f(z) = 1$ for all z in X, then f is certainly $\mathbf{Z}$-approximable on X. Also, the function

$$f'(z) = |z_1|, \qquad z \in h(X),$$

is an extension of f to $h(X)$ which is continuous on $h(X)$ and holomorphic on $h(X)^\circ$ (indeed, $h(X)^\circ$ is empty). But it is clear that the unique extension $\tilde{f}$ of f which is Z-approximable on $h(X)$ is simply $\tilde{f}(z) = 1$, $z \in h(X)$.

In view of the foregoing argument we need only consider the problem for those X such that $X = h(X)$, i.e., the polynomially convex subsets of $\mathbf{C}^n$.

PROPOSITION 7.10. *Let X be a compact subset of $\mathbf{C}^n$ and f a complex valued function on X. If f is R-approximable on X then the coefficients of its power series expansion about each point in $R^n \cap X^\circ$ are elements of R.*

PROOF. Since f is R-approximable on X and $X^\circ \subset h(X)^\circ$, Proposition 7.9 implies that f has a power series expansion (with nonzero radius of convergence) about each point of X°. For z_0 in $R^n \cap X^\circ$ choose $\delta > 0$ such that the polydisk $z_0 + \delta D^n \subset X^\circ$ and

$$f(z) = \sum_{k \in \mathbf{N}^n} b_k (z - z_0)^k, \qquad z \in (z_0 + \delta D^n).$$

If (p_m) is a sequence of polynomials in $R[z]$ converging uniformly on X to f we have $p_m(z + z_0) \to f(z + z_0)$ uniformly on δD^n and

$$f(z + z_0) = \sum_{k \in \mathbf{N}^n} b_k z^k, \qquad z \in \delta D^n.$$

If, for each m

$$p_m(z + z_0) = \sum_{k \in \mathbf{N}^n} a_k^{(m)} z^k$$

where, of course, all but finitely many of the $a_k^{(m)}$, for fixed m, are 0, we have $a_k^{(m)} \in R$ since $z_0 \in R^n$ and the coefficients of p_m are in R by hypothesis. Since the convergence of $\{p_m(z + z_0)\}$ to $f(z + z_0)$ is uniform on δD^n, we have that the respective power series converge term by term. Thus, for each $k \in \mathbf{N}^n$, $a_k^{(m)} \to b_k$ as $m \to \infty$. But R is a discrete subgroup (additive) of $\mathbf{C}$, hence closed (Hewitt and Ross [**63**, (5.10)]) whence $b_k \in R$ for each k. $\square$

For completeness we quote the following. The proof is the same as in Theorem 2.11.

PROPOSITION 7.11. *Let X be a compact subset of $\mathbf{C}$ with $d(X) \geqslant 1$. Then a complex valued function f on X is R-approximable on X if and only if it is already an element of $R[z]$.*

By restricting a complex valued function on $\mathbf{C}^n$ to a function of one complex variable, in certain ways, we can get additional necessary conditions for approximability as follows.

For any pair (k, a) such that $(1 \leqslant k \leqslant n)$ and $a \in R^{n-1}$ define an injection $j_{k,a}\colon \mathbf{C} \to \mathbf{C}^n$ by

$$j_{k,a}(z) = (a_1, \ldots, a_{k-1}, z, a_k, \ldots, a_{n-1}).$$

Then $\pi_k \circ j_{k,a} = \mathrm{id}_{\mathbf{C}}$ where π_k is the kth coordinate projection of $\mathbf{C}^n$, $\mathrm{id}_{\mathbf{C}}$ is the identity map of $\mathbf{C}$, and $\circ$ is the usual composition of functions operation. For each pair (k, a) as above let

$$X_{k,a} = \pi_k(X \cap j_{k,a}(\mathbf{C})).$$

If f is a complex valued function on X and $\|f - p\|_X < \varepsilon$ where $p \in R[z_1, \ldots, z_n]$ we have $\|f \circ j_{k,a} - p \circ j_{k,a}\|_{X_{k,a}} < \varepsilon$ and $p \circ j_{k,a} \in R[z]$ $(z \in \mathbf{C})$. Thus in order that f be R-approximable on X, it is necessary that for all (k, a) as above, $f \circ j_{k,a}$ be R-approximable on $X_{k,a}$, i.e., a number of one-dimensional problems arise and Chapter 4 applies to these. It is clear from the discreteness of R^n that only finitely many of the $X_{k,a}$ are nonempty.

We turn now to sufficient conditions for approximability. In the remainder of this chapter we take the ring of coefficients of the approximating polynomials to be any discrete subring A of $\mathbf{C}$ with rank 2. We also always take X to be polynomially convex.

PROPOSITION 7.12. *If z' is an element of A^n then a complex valued function $f(z)$ is A-approximable on a subset X of $\mathbf{C}^n$ if and only if $f(z - z')$ is A-approximable on $z' + X$.*

The proof is obvious since if $q(z) \in A[z]$ then so is $q(z - z')$.

The following definition partially delineates the class of sets to which the main theorem of this section applies.

DEFINITION 7.13. A compact subset X of $\mathbf{C}^n$ is said to be Mergelyan if it satisfies the following condition. Any complex valued function which is continuous on X and holomorphic on X° is $\mathbf{C}$-approximable on X.

The terminology is motivated by the following theorem (Mergelyan **[51]**, Rudin **[74]**).

THEOREM 7.14. *A compact subset X of $\mathbf{C}$ is Mergelyan if and only if its complement $\mathbf{C} \setminus X$ is connected.*

By Proposition 7.4 we see that an equivalent condition is that X be polynomially convex, at least in case $X \subset \mathbf{C}$. The answer to the question of an equivalent condition for compact subsets of $\mathbf{C}^n$, $n > 1$, is not known to the writer, but is certainly more complicated. Indeed, we can see that Lavrent'ev's theorem (Proposition 4.2), which is an obvious consequence of Mergelyan's theorem above, fails if we attempt to apply it word for word to compact subsets of $\mathbf{C}^n$, $n > 1$, as follows. Let X be as in Example 7.8. Then $h(X)$ has void interior since $n > 1$, and its complement $\mathbf{C}^n \setminus X$ is connected as can be seen by elementary means. Thus $h(X)$ satisfies the hypotheses of Mergelyan's theorem except that $X \not\subset \mathbf{C}$. The continuous function $f'(z) = |z_1|$, $z \in h(X)$, is not $\mathbf{C}$-approximable on $h(X)$ since by Proposition 7.3 there exists exactly one such approximable extension of $f(z) = 1$, $z \in X$, to $h(X)$ and this is obviously the function $\tilde{f}(z) = 1$, $z \in h(X)$. Thus neither Lavrent'ev's nor Mergelyan's theorem can be extended to higher dimensions.

A theorem which does extend, in a sense, is Runge's theorem. In one of its forms this theorem states that every complex valued function holomorphic in a neighborhood of a compact set X of $\mathbf{C}$ is $\mathbf{C}$-approximable on X if and only if $\mathbf{C} \setminus X$ is connected (cf. Rudin [**74**]). From Proposition 7.4 we know that $\mathbf{C} \setminus X$ is connected if and only if X is polynomially convex. It is the last criterion which remains in force as we pass to higher dimensions.

PROPOSITION 7.15. *Let X be a polynomially convex subset of $\mathbf{C}^n$. Then any function which is holomorphic in a neighborhood of X is $\mathbf{C}$-approximable on X.*

For a proof see Hörmander [**66**, Theorem 2.7.7].

From Proposition 7.15 it is easy to see that every Mergelyan set is polynomially convex as follows. We assume that X is Mergelyan but not polynomially convex and derive a contradiction. There exists $z_0 \notin X$ such that

$$|p(z_0)| \leqslant \|p\|_X, \qquad p \in \mathbf{C}[z]. \tag{$*$}$$

Define f on $X \cup \{z_0\}$ by $f[X] = \{0\}$ and $f(z_0) = 1$. Since X is Mergelyan there is a sequence $\{p_n\}$ in $\mathbf{C}[z]$ which converges uniformly to f on X. By Proposition 7.3 (proof) p_n converges uniformly to f on $h(X)$. Since $z_0 \in h(X)$ we have that $p_n(z_0) \to f(z_0) = 1$ which contradicts $(*)$ since $\|p_n\|_X \to 0$.

The following technical result will be used to exhibit a large class of Mergelyan subsets of $\mathbf{C}^n$, for any n.

PROPOSITION 7.16. *Let X_1 and X_2 be disjoint compact subsets of C such that $X_1 \cup X_2$ is polynomially convex and each X_i satisfies the following condition $(*)$: There exist z_i' in X_i and $\delta_i > 0$ such that whenever $1 < a < 1 + \delta_i$ we have*

$$(aX_i - (a-1)z_i')^\circ \supset X_i.$$

Then $X_1 \cup X_2$ is Mergelyan. In addition, any polynomially convex compact subset of $\mathbf{C}^n$ which satisfies condition $(*)$ is Mergelyan.

Condition $(*)$ might also be expressed by saying that the dilation of X_i about z_i' of magnitude a is a neighborhood of X for all a in $(1, 1 + \delta_i)$. The proof is rather complicated in its details, but the idea is simply to dilate each X_i slightly so as to obtain from the original function one which is holomorphic in a neighborhood of $X_1 \cup X_2$ and then apply Proposition 7.15.

PROOF. Let f be continuous on $X_1 \cup X_2$, holomorphic on $(X_1 \cup X_2)^\circ$ and $\varepsilon > 0$. For any positive number a, define the maps $\Delta_a^{(i)}: X_i \to \Delta_a^{(i)}(X_i) \subset \mathbf{C}^n$, $i = 1, 2$, by

$$\Delta_a^{(i)}(z) = az - (a-1)z_i'.$$

It is easy to see that

$$\left(\Delta_a^{(i)}\right)^{-1}(z) = a^{-1}z - (a^{-1} - 1)z_i'$$

so that $\Delta_a^{(i)}$ is a homeomorphism for each i.

Since $X_1 \cup X_2$ is compact, f is uniformly continuous on $X_1 \cup X_2$ and there exists $\tilde{\delta} > 0$ such that $z_1, z_2 \in X_1 \cup X_2$ and $|z_1 - z_2| < \tilde{\delta}$ imply $|f(z_1) - f(z_2)|$

$< \varepsilon/2$. Also by the compactness of X_1 and X_2, $d = \inf|X_1 - X_2| > 0$ where the equality serves to define d. It is easy to see that for each i

$$\|\Delta_a^{(i)}(z) - z\|_{X_i} = |1 - a| \, \|z - z_i'\|_{X_i};$$

hence we can fix $a > 1$ such that

$$\|\Delta_{a^{-1}}^{(i)}(z) - z\|_{X_i} < \tilde{\delta}, \qquad i = 1, 2, \tag{1}$$

$$\|\Delta_a^{(i)}(z) - z\|_{X_i} < d/2, \qquad i = 1, 2, \tag{2}$$

$a < \delta_1$ and $a < \delta_2$. Then $\Delta_a(X_i)$ is a neighborhood of X_i by hypothesis ($i = 1, 2$) and by (2)

$$\left(\Delta_a^{(1)}[X_1]\right) \cap \left(\Delta_a^{(2)}[X_2]\right) = \varnothing.$$

Thus we can define a function $\tilde{f}$ on $\tilde{X} = (\Delta_a^{(1)}[X_1]) \cup (\Delta_a^{(2)}[X_2])$ by

$$\tilde{f}|_{\Delta_a^{(i)} X_i} = f \circ \Delta_{a^{-1}}^{(i)}.$$

To show that $\tilde{f}$ is holomorphic on the neighborhood $(\Delta_a^{(1)}[X_1])^\circ \cup (\Delta_a^{(2)}[X_2])^\circ$ of $X_1 \cup X_2$, it suffices to show that $\tilde{f}|_{\Delta_a^{(i)}[X_i]}$ is holomorphic on $(\Delta_a^{(i)}[X_i])^\circ$, $i = 1, 2$. But $(\Delta_a^{(i)})^{-1}[(\Delta_a^{(i)}[X_i])^\circ]$ is an open subset of X_i by the continuity of $\Delta_a^{(i)}$ and $\Delta_{a^{-1}}^{(i)} = (\Delta_a^{(i)})^{-1}$. Since $\tilde{f}|_{\Delta_a^{(i)}[X_i]}$ is the composition of two holomorphic functions, it too is holomorphic. Thus $\tilde{f}$ is holomorphic on a neighborhood of $X_1 \cup X_2$. By hypothesis $X_1 \cup X_2$ is polynomially convex; hence, by Proposition 7.15, $\tilde{f}$ is **C**-approximable on $\tilde{X}$. Let $p \in \mathbf{C}[z]$ with

$$\|\tilde{f} - p\|_{X_1 \cup X_2} < \varepsilon/2. \tag{3}$$

We claim that $\|f - p\|_{X_1 \cup X_2} < \varepsilon$ which follows from (3) and the fact that for $i = 1$ and 2

$$|f(z) - f(\Delta_{a^{-1}}^{(i)}(z))| < \varepsilon/2, \qquad z \in X_i.$$

The latter follows from (1) and the choice of $\tilde{\delta}$. Thus f is **C**-approximable on $X_1 \cup X_2$ which proves the first statement in the proposition.

It is easy to see how to use some of the same ideas to prove the last statement of the proposition. □

We have no need for Proposition 7.16 in greater generality but we note in passing that the first part of the proposition is true for finitely many X_i as can be seen immediately from the proof.

We have seen that every convex compact set is polynomially convex. If the set also contains an interior point, then by the following corollary it is Mergelyan as well.

It is easy to see that the assumption of the interior point is necessary (if $n > 1$) by the following example. Let $X = \{z \in \mathbf{C}^n\colon |z_1| = 1\}$. Then applying the maximum modulus principle in the single variable z_1 we easily see that $h(X) = \{z\colon |z_1| \leqslant 1\}$ (cf. Example 7.8). Also $(h(X))^\circ = \varnothing$ $(n > 1)$ so the function $f(z) = |z_1|$ is continuous on $h(X)$ and analytic on $(h(X))^\circ$. It is not approximable, however, since by Proposition 7.9 there exists a unique approximable extension of f from X to $h(X)$ and this is obviously the function $h(z) \equiv 1$.

COROLLARY 7.17. *Every compact convex subset X of $\mathbf{C}^n$ with nonvoid interior is Mergelyan.*

PROOF. Let z' be an interior point of X. Since the criteria in the statement of the result are invariant under translation, we can assume $z' = 0$. We know from Example 7.2 that the convexity of X implies that it is polynomially convex. Thus by Proposition 7.16 it suffices to show that $(aX)^\circ \supset X$ for every $a > 1$. But given z in X we see that $z \in (aX)^\circ$ as follows. Since 0 is interior to X, there is an open neighborhood V of 0 with $V \subset X$. Since aX is convex and $V \subset X \subset aX$ we have

$$(1 - 1/a)V + (1/a)(az) \subset aX. \tag{1}$$

Now notice that the left-hand side of (1) is an open set containing z. □

EXAMPLE 7.18. It is clear that every set of the form

$$X = \{z \in \mathbf{C}^n: |z - z_1'| \leqslant \rho_1, \ldots, |z - z_n'| \leqslant \rho_n\}$$

satisfies the hypothesis of Corollary 7.17, hence is Mergelyan, where $z' = (z_1', \ldots, z_n')$ is any point in $\mathbf{C}^n$ and the ρ_i's are any positive numbers. Such a set is called a *closed polydisk* with center z' and multiradius $(\rho_1, \ldots, \rho_n)$. The similarly defined set with the n inequalities replaced by strict inequalities is called an *open polydisk*.

Similarly the class of Mergelyan sets contains the sets of the form $X = \{z \in \mathbf{C}^n: |z - z'| \leqslant \rho\}$. Such a set X is called a closed ball of radius ρ and centered at z'.

At this point one might conjecture that every finite union of disjoint compact sets satisfying condition $(*)$ of Proposition 7.16 is Mergelyan. Actually, the polynomial convexity of the union is necessary and sufficient for this.

The necessity was seen in the comment following Proposition 7.15 and the sufficiency follows from the comments following Proposition 7.16. The conjecture is false, however, even for convex sets with interior, in view of the example by Kallin **[65]** of three disjoint polydisks whose union is not polynomially convex.

The following is the main result on approximation by integral polynomials in $\mathbf{C}^n$ where $n > 1$. For $n = 1$ a much more complete result is known (Theorem 7.21).

THEOREM 7.19. *Let A be a discrete subring of $\mathbf{C}$ with rank 2 and let X be a Mergelyan subset of the open unit polydisk, i.e., $X \subset \{z \in \mathbf{C}^n: |z_1| < 1, \ldots, |z_n| < 1\}$, and 0 in X°. If f is a complex valued function on X, the following are equivalent*:

(i) *The function f is A-approximable on X.*

(ii) *The function f is continuous on X, holomorphic on X°, and the coefficients of its power series expansion about 0 lie in A.*

PROOF. By Propositions 7.9 and 7.10 we have that (i) implies (ii).

Conversely, assume that (ii) holds and that ε is any positive number. Since X

is compact and the coordinate projections are continuous, we have, where ρ and the ρ_i's are so defined, $\rho_i = \|z_i\|_X < 1$ and so $\rho = \max_{1 \leq i \leq n} \rho_i < 1$. It is easy to see that

$$\left(\frac{1}{1-\rho}\right)^n = \sum_{k \in \mathbf{N}^n} \rho^{k_1 + \cdots + k_n}.$$

Let $\delta > 0$ be chosen as in Proposition 1.2. There exists a finite subset F of $\mathbf{N}^n$ such that $\sum_{k \in \mathbf{N}^n \setminus F} \rho^{k_1 + \cdots + k_n} < \varepsilon/3\delta$. Since X is Mergelyan by hypothesis, there exists a sequence of polynomials $\{p_m\}$ in $\mathbf{C}[z]$ such that $p_m \to f$ uniformly on X. Let $p_m(z) = \sum_{k \in \mathbf{N}^n} a_k^{(m)} z^k$ where, of course, all but finitely many of the $a_k^{(m)} = 0$, for each m, and let $f(z) = \sum_{k \in \mathbf{N}^n} a_k z^k$ in a neighborhood of the origin. Then, as noted in the proof of Proposition 7.10, for each k in $\mathbf{N}^n$, $a_k^{(m)} \to a_k$ as $m \to \infty$. Thus there exists a positive integer N such that $m > N$ implies

$$\|p_m - f\|_X < \varepsilon/3 \tag{$*$}$$

and

$$|a_k^{(m)} - a_k| < \varepsilon/3M(\operatorname{card} F), \qquad k \in F,$$

where $M = \max_{k \in F} \|z^k\|_X$. Thus if $m > N$ and $[p_m]$ denotes the polynomial p_m with each coefficient replaced by a nearest element of A, we have

$$\begin{aligned} \|p_m - [p_m]\| &\leq \sum_{k \in F} |a_k^{(m)} - a_k| \, \|z^k\| + \sum_{k \notin F} \delta \|z^k\| \\ &\leq \sum_{k \in F} \frac{\varepsilon}{3M(\operatorname{card} F)} M + \delta \sum_{k \notin F} \rho^{k_1 + \cdots + k_n} \\ &< \varepsilon/3 + \varepsilon/3 = 2\varepsilon/3. \end{aligned}$$

This estimate together with $(*)$ gives $\|[p_n] - f\|_X < \varepsilon$. □

The problem is, in the sense given by Proposition 7.12, invariant under translation by elements of A^n and so the theorem could have been stated for Mergelyan subsets of the open unit polydisk centered at any element of A^n.

A similar result to that in Theorem 7.19 holds when X does not contain the origin, as follows.

COROLLARY 7.20. *Let A be a discrete subring of $\mathbf{C}$ with rank 2 and let X be a polynomially convex compact subset of the open unit polydisk, $0 \notin X$, and satisfying condition $(*)$ of Proposition 7.16. Then every continuous complex valued function on X which is holomorphic on $X°$ is A-approximable on X.*

PROOF. We first establish the "separation lemma" of Kallin [65] which says that if X_1 and X_2 are two compact subsets of $\mathbf{C}^n$ and there exists a polynomial f such that $h(f(X_1)) \cap h(f(X_2)) = 0$, then $h(X_1 \cup X_2) = h(X_1) \cup h(X_2)$. The inclusion $h(X_1 \cup X_2) \supset h(X_1) \cup h(X_2)$ is obvious from the definition of a polynomially convex hull; hence we need only establish the reverse inclusion. To this end suppose that $y \notin h(X_1) \cup h(X_2)$. There are two cases. First suppose that $f(y) \notin h(f(X_1)) \cup (f(X_2))$. Then $f(y) \notin h(f(X_1))$; hence by Runge's theorem there is a polynomial p_1 in one variable such that $p_1(f(y))$ is near 1 and $p_1 \cdot f$ is near 0 on X_1. We argue the same way with the subscript 1 replaced by 2 and then $(p_1 \cdot p_2)(f(y))$ is near 1 but $(p_1 \cdot p_2) \cdot f$ is near 0 on $X_1 \cup X_2$ which shows that $y \notin h(X_1 \cup X_2)$. For the other case we suppose that $f(y) \in h(f(X_1)) \cup h(f(X_2))$. We suppose without loss of generality that $f(y) \in h(f(X_1))$. Since $y \notin h(X_1)$ there exists a polynomial s with $s(y) = 1$ and $|s| < \frac{1}{2}$ on X_1. Let $M = \sup|s(X_2)|$. By Runge's thereom there exists a polynomial r in one variable with $|r - 1| < \frac{1}{3}$ on $f(X_1)$ and $|r| < 1/(2M)$ on $f(X_2)$. Then $g(r \cdot f) \cdot s$ is a polynomial satisfying $|g(y) - 1| < \frac{1}{3}$ and $|g| < \frac{2}{3}$ on $X_1 \cup X_2$. Thus $y \notin h(X_1 \cup X_2)$ and Kallin's separation lemma is proved.

Now suppose we are in the situation stated in the corollary. Since $0 \notin X$ and X is polynomially convex by hypothesis, there is a p in $\mathbf{C}[z]$ with $|p(0)| > \|p\|_X$. Thus $p(0)$ is a point of the complex w-plane lying outside of the circle $|w| = \|p\|_X$ which in turn contains $p(X)$. By the continuity of p there exists a closed ball X_1 centered at $z = 0$ such that $p(X_1)$ lies outside of the circle $|w| = \|p\|_X$. By Proposition 7.4 we have

$$h(p(X_1)) \cap h(p(X)) = \varnothing$$

and applying Kallin's separation lemma we see that $X_1 \cup X$ is polynomially convex. It is clear that every dilation of X_1 about the origin of magnitude $a > 1$ is a neighborhood of X_1. Thus X_1 and X satisfy the hypotheses of Proposition 7.16; hence $X_1 \cup X$ is Mergelyan. Let $f \in C(X)$ and f be holomorphic on X°. Extend f to $X_1 \cup X$ by setting $f(X_1) = \{0\}$. Then $f \in C(X_1 \cup X)$, f is holomorphic on $(X_1 \cup X)^\circ = X_1^\circ \cup X^\circ$, and the coefficients of the power series expansion of f about 0 are all 0, hence lie in A. Thus f is A-approximable on $X_1 \cup X$ by Theorem 7.19. □

In the case of dimension 1 the problem has a complete solution similar to that in Chapter 4. A new condition enters at the points of X° which are algebraic integers and whose conjugates over L are also in X° (where X° denotes the interior of X). From the results at the beginning of this chapter we see that we can restrict our attention to the case where X is Mergelyan and $d(X) < 1$.

THEOREM 7.21. *Let A be a discrete subring of $\mathbf{C}$ with rank 2 and containing the identity and X a Mergelyan subset of $\mathbf{C}$ with $d(X) < 1$. A complex valued function f on X is A-approximable on X if and only if it is continuous on X, holomorphic on X° and satisfies the following condition. If m is any positive integer, then there exists a q in $A[z]$ such that*

$$\begin{aligned} q(z) &= f(z), \qquad z \in J_0(X); \\ q^{(\nu)}(z) &= f^{(\nu)}(z), \qquad z \in J_0(X^\circ), \quad 1 \leqslant \nu \leqslant m. \end{aligned} \tag{1}$$

PROOF. By Theorem 7.14 the complement of X in $\mathbf{C}$ has no bounded components, so by Proposition 7.4, $X = h(X)$. Then Proposition 7.9 shows that for f to be A-approximable, it is necessary that it be continuous on X and holomorphic on X°. The necessity of the remaining condition on f is given by Proposition 3.10.

The sufficiency of the conditions is established with somewhat more difficulty as follows.

Let $A \subset I_L$ where L is the unique imaginary quadratic field such that $A \subset I_L$, as usual. Since A contains the identity, there exists a monic $\tilde{q}$ in $A[z]$ with $\|\tilde{q}\| < 1$ by Lemma 4.3. For any $\varepsilon_0 > 0$ there exists a positive integer κ such that

$$\|\tilde{q}\|_X^\kappa / (1 - \|\tilde{q}\|_X) < \varepsilon_0. \tag{2}$$

Obviously it suffices to approximate the difference between f and any element of $A[z]$. We will use this fact repeatedly. Thus, by (1), we may assume

$$f(z) = 0, \qquad z \in J_0(X),$$
$$f^{(\nu)}(z) = 0, \qquad z \in J_0(X^\circ), \quad 1 \leqslant \nu \leqslant m,$$

where m is taken to be the product of κ with the maximum of the multiplicities of the roots of $\tilde{q}$. If we let q_1 be the product of the minimal polynomials associated with the complete sets of conjugate (over L) algebraic integers making up $J_0(X^\circ)$ and $J_0(\partial X)$ where ∂X denotes the boundary of X, then $m_0 q_1 \in A[z]$ where m_0 is the positive integer in Proposition 1.11 such that $m_0 I_L \subset A$. Let $Z_{\tilde{q}}$ denote the set of roots of $\tilde{q}$. Since $\tilde{q}$ is a monic polynomial in $I_L[z]$, its roots are all algebraic integers. By Lemma 3.6, $J_0(X^\circ) \subset Z_{\tilde{q}}$. The elements of $(Z_{\tilde{q}} \cap X^\circ) \setminus J_0(X^\circ)$ have the property that at least one of their conjugates (over L) lies outside of X°. Thus by Theorem A.5 there is a q_2 in $A[z]$ such that

$$\left|f^{(\nu)}(z) - ((m_0 q_1)^m q_2)^{(\nu)}(z)\right| < \varepsilon_1, \qquad z \in (Z_{\tilde{q}} \cap X^\circ) \setminus J_0(X^\circ), \quad 0 \leqslant \nu \leqslant m,$$

where ε_1 is any preassigned positive number. From this we have that

$$|f(z) - ((m_0 q_1)^m q_2)(z)| < \varepsilon_1, \qquad z \in Z_{\tilde{q}} \cap X;$$
$$\left|f^{(\nu)}(z) - ((m_0 q_1)^m q_2)^{(\nu)}(z)\right| < \varepsilon_1, \qquad z \in Z_{\tilde{q}} \cap X^\circ, \quad 1 \leqslant \nu \leqslant m, \tag{3}$$

since at the points newly introduced the two terms in the left-hand side of the relevant inequality in (3) are both zero. Thus without loss of generality we can assume

$$|f(z)| < \varepsilon_1, \qquad z \in Z_{\tilde{q}} \cap X, \text{ and}$$
$$|f^{(\nu)}(z)| < \varepsilon_1, \qquad z \in Z_{\tilde{q}} \cap X^\circ, \quad 1 \leqslant \nu \leqslant m.$$

Let p be the Hermite interpolation polynomial defined by

$$p(z) = f(z), \qquad z \in Z_{\tilde{q}} \cap X,$$
$$p^{(\nu)}(z) = f^{(\nu)}(z), \qquad z \in Z_{\tilde{q}} \cap X^\circ, \quad 1 \leqslant \nu \leqslant m.$$

Then we have, by a slight generalization of Lemma 1, Chapter XI of Walsh **[60]**

$$\|p\|_X \leqslant \varepsilon_1 A_0(\varepsilon_0) \tag{4}$$

where $A_0(\varepsilon_0)$ depends only on X, the points of interpolation, and their multiplicities–hence only on ε_0. Factor $\tilde{q}$ in the form $\tilde{q} = p_1 p_2 p_3$ where the roots of p_1 lie in X°, the roots of p_2 lie in ∂X and the roots of p_3 lie outside of X. Then the function $(f-p)/p_1^\kappa p_3^\kappa$ is continuous on X and holomorphic on X°. Since X is Mergelyan, for any $\varepsilon_2 > 0$ there exists a polynomial R_0 such that

$$\|(f-p)/p_1^\kappa p_3^\kappa - R_0\|_X < \varepsilon_2;$$

hence

$$\|f - p - p_1^\kappa p_3^\kappa R_0\|_X < \varepsilon_2 A_1(\varepsilon_0) \tag{5}$$

where $A_1(\varepsilon_0) = \|p_1^\kappa p_3^\kappa\|_X$. Also, by Walsh [60, Chapter 11, Theorem 1, Corollary], we can assume that $R_0(z) = 0$ at all points z in $Z_{\tilde{q}} \cap \partial X$ since $f - p$ is zero at these points. Thus $R_0 = \tilde{p}_2 R_1$ where $\tilde{p}_2$ is the polynomial with the same roots as p_2 but only simple roots and R_1 is some polynomial. Since the roots of $\tilde{p}_2$ all lie on ∂X and X has a connected complement (Theorem 7.14), there is a branch of $\tilde{p}_2^{1/m}$ which is holomorphic on X° and continuous on X. Thus from the hypothesis that X is Mergelyan and the same theorem of Walsh, there exists a polynomial R_2 such that

$$\|\tilde{p}_2^{1/m} - \tilde{p}_2 R_2\|_X < \varepsilon_3$$

where ε_3 is any preassigned positive number and $\varepsilon_3 < 1$. But for any complex numbers a and b

$$|a^m - b^m| \leqslant |a-b|(|a| + |b|)^{m-1};$$

hence

$$\|\tilde{p}_2 - \tilde{p}_2^m R_2^m\|_X < \varepsilon_3(A_2(\varepsilon_0) + \varepsilon_3)^{m-1} < \varepsilon_3(A_2(\varepsilon_0) + 1)^{m-1} = \varepsilon_3 A_3(\varepsilon_0)$$

where the equality serves to define $A_3(\varepsilon_0)$ and we have used the estimate

$$\begin{aligned}|\tilde{p}_2^{1/m}| + |\tilde{p}_2 R_2| &= |\tilde{p}_2^{1/m}| + |\tilde{p}_2 R_2 - \tilde{p}_2^{1/m} + \tilde{p}_2^{1/m}| \\ &\leqslant |\tilde{p}_2^{1/m}| + |\tilde{p}_2 R_2 - \tilde{p}_2^{1/m}| + |\tilde{p}_2^{1/m}| \\ &= 2|\tilde{p}_2^{1/m}| + |\tilde{p}_2 R_2 - \tilde{p}_2^{1/m}| \\ &< A_2(\varepsilon_0) + \varepsilon_3.\end{aligned}$$

Set $A_4(\varepsilon_0, \varepsilon_2) = \|p_1^m p_3^m R_1\|$. Then by (5) we have

$$\|f - p - p_1^\kappa p_3^\kappa \tilde{p}_2 R_2^\kappa R_1\| < \varepsilon_2 A_1(\varepsilon_0) + \varepsilon_3 A_4(\varepsilon_0, \varepsilon_2) A_3(\varepsilon_0).$$

By our choice of m we can write $\tilde{p}_2^m = p_2^\kappa R_3$ for some polynomial R_3 and setting $R_4 = R_3 R_2^\kappa R_1$ we have

$$\begin{aligned}\|f - p - \tilde{q}^\kappa R_4\|_X &= \|f - p - p_1^\kappa p_2^\kappa p_3^\kappa R_4\|_X \\ &< \varepsilon_2 A_1(\varepsilon_0) + \varepsilon_3 A_4(\varepsilon_0, \varepsilon_2) A_3(\varepsilon_0).\end{aligned}$$

From Lemma 4.4 we see that we can represent R_4 as a (finite) sum of the form $R_4 = \Sigma_{j \geqslant 0} h_j \tilde{q}^j$ where each h_j is a polynomial and $\deg h_j < \deg \tilde{q}$. Thus

$$\tilde{q}^\kappa R_4 = \sum_{j \geqslant \kappa} h_{j-\kappa} \tilde{q}^j.$$

Denote by $[h_{j-\kappa}]$ the polynomial obtained from $h_{j-\kappa}$ by replacing each coefficient of $h_{j-\kappa}$ by a nearest element of A. If we set $[\tilde{q}^\kappa R_4] = \Sigma_{j \geqslant \kappa}[h_{j-\kappa}]\tilde{q}^j$ then $[\tilde{q}^\kappa R_4]$ is an element of $A[z]$ and

$$\|\tilde{q}^{\kappa}R_4 - [\tilde{q}^{\kappa}R_4]\|_X \leqslant \Big(\delta(\deg \tilde{q}) \max_{0 \leqslant s \leqslant \deg \tilde{q}} \|x^s\|\Big) \sum_{j \geqslant \kappa} \|\tilde{q}\|^j$$

$$= \tilde{A} \sum_{j \geqslant \kappa} \|\tilde{q}\|^j \leqslant \tilde{A} \frac{\|\tilde{q}\|^{\kappa}}{1 - \|\tilde{q}\|} < \varepsilon_0 \tilde{A}.$$

Thus

$$\|f - p - [\tilde{q}^{\kappa}R_4]\| \leqslant \varepsilon_0\tilde{A} + \varepsilon_2 A_1(\varepsilon_0) + \varepsilon_3 A_4(\varepsilon_0, \varepsilon_2) A_3(\varepsilon_0)$$

and, finally, by (4),

$$\|f - [\tilde{q}^{\kappa}R_4]\| \leqslant \varepsilon_0\tilde{A} + \varepsilon_1 A_0(\varepsilon_0) + \varepsilon_2 A_1(\varepsilon_0) + \varepsilon_3 A_4(\varepsilon_0, \varepsilon_2) A_3(\varepsilon_0).$$

This proves the sufficiency of our conditions since we can first select ε_0 to make $\varepsilon_0\tilde{A}$ arbitrarily small, then ε_1 to make $\varepsilon_1 A_0(\varepsilon_0)$ arbitrarily small, etc. □

COROLLARY 7.22. *If $A = I_L$ then the last condition in the theorem can be replaced by the condition that the Hermite interpolation polynomials determined by* (1) *have integral coefficients* (*i.e., lie in $I_L[z]$*).

PROOF. The new condition obviously implies the old one. Conversely, let q satisfy the conditions in (1). Define w to be a certain product of the minimal polynomials associated with $J_0(X)$ as follows. The minimal polynomials associated with $J_0(X^\circ)$ enter with the power m and the other minimal polynomials enter only to the first power. Then w is a monic integral polynomial. Since q is also an integral polynomial (i.e., in $I_L[z]$), a cursory inspection of the usual division algorithm shows that

$$q = pw + \tilde{q}, \qquad \deg \tilde{q} < \deg w,$$

with $\tilde{q}$ an integral polynomial also. From this equation, the bound on the degree of $\tilde{q}$, and the uniqueness of Hermite interpolation polynomials we see that $\tilde{q}$ is this interpolation polynomial. □

In the historical notes (Part IV) there is an example showing that we cannot simply drop the condition $A = I_L$ from the hypotheses of this corollary.

An example of the use of these results follows.

EXAMPLE 7.23. Let X be a circle of radius $r < 1$ and centered at the origin of the complex plane. There is no loss of generality in the assumption $r < 1$ since, by Example 2.14, $d(X) = r$ and if $d(X) \geqslant 1$ the problem is trivial by Proposition 7.11. Let f be a continuous complex valued function defined on X and A a discrete subring of **C** of rank 2. By Proposition 7.4, $h(X) = rD$, where D is the closed unit disk. By Proposition 7.9, in order that f be A-approximable on X, it is necessary that f have a continuous extension to $h(X)$ which is holomorphic on $h(X)^\circ$ and A-approximable on $h(X)$.

It is well known that a continuous function g defined on $|z| = 1$ has an extension which is continuous on $|z| \leqslant 1$ and holomorphic on $|z| < 1$ if and only if the Fourier coefficients $\hat{g}(n)$ of the function are all zero for $n < 0$ (Hoffman [**62**, p. 42]). It is easy to see from this that f has a continuous extension $\tilde{f}$ to $h(X)$ which is holomorphic on $h(X)^\circ$ if and only if

$$\int_{|z|=1} f(rz)z^m\,dz = 0, \qquad m = 0, 1, \ldots .$$

By a change of variable this is equivalent to

$$r^{-(m+1)}\int_X f(z)z^m\,dz = 0, \qquad m = 0, 1, \ldots .$$

It is clear from the maximum modulus principle that $\tilde{f}$ is uniquely determined, if it exists. By Proposition 7.10, if f is to be A-approximable on X, then the coefficients of the power series expansion of $\tilde{f}$ about 0 must be in A, that is,

$$\frac{1}{2\pi i}\int_X f(z)z^{-(m+1)}\,dz \in A, \qquad m = 0, 1, \ldots .$$

In summary then, in order that f be A-approximable, it is necessary that

$$\frac{1}{2\pi i}\int_X f(z)z^{m-1}\,dz \begin{cases} = 0 & \text{if } m = 1, 2, \ldots, \\ \in A & \text{if } m = 0, -1, \ldots . \end{cases}$$

These conditions are also sufficient by the preceding discussion and Theorem 7.19.

In the one dimensional case, if the function to be approximated is rational and the ring of coefficients A is a unique factorization domain, the problem has a particularly simple solution as follows (Ferguson [**69**]). We assume for the remainder of this chapter that X is a Mergelyan subset of the open unit disk D° with 0 in X°. If $0 \notin X$ we could simply add a sufficiently small neighborhood of 0 to X and extend f by setting it identically zero on the neighborhood. The new set X would still be Mergelyan by Theorem 7.14.

THEOREM 7.24. *A rational function f is A-approximable on X if and only if it can be represented in the form $f = p/q$ where p and q are in $A[z]$, $q(0)$ is a unit of A, and the roots of q lie outside of X.*

PROOF. First suppose that f is represented as in the statement of the theorem. Then f is continuous on X and holomorphic on X°; so by Theorem 7.19 it suffices to prove that the coefficients of the power series expansion

$$f(z) = \sum_{k=0}^{\infty} c_k z^k \tag{1}$$

lie in A. Let

$$p(z) = \sum_{k=0}^{n} a_k z^k \tag{2}$$

and

$$q(z) = \sum_{k=0}^{n} b_k z^k. \tag{3}$$

Then (1) is equivalent to the infinite system of equations

$$b_0c_0 = a_0, \tag{4}$$

$$b_0c_1 + b_1c_0 = a_1, \tag{5}$$

$$\vdots$$

From (4), since $b_0 = q(0)$ has an inverse in A, we see that $c_0 \in A$. Then from (5) we see that $c_1 \in A$ and so on.

Conversely, suppose that f is a rational function, (1) holds in a neighborhood of the origin, and f is A-approximable on X. Since f is rational, $f = p/q$ for p and q polynomials with complex coefficients given, say, by (2) and (3) respectively. The roots of q lie outside of X and by Proposition 7.10 the c_k's are elements of A. By (1) the infinite system of equations

$$b_0c_k + b_1c_{k-1} + \cdots + b_nc_{k-n} = 0 \qquad (k > n) \tag{6}$$

is solvable for $b_0, \ldots, b_n$ not all zero. Since the c_k's are in A, the b_k's are in the field of quotients of A and, since the equations are homogeneous, we can assume that the b_k's are in A. Now define the a_k's to equal the left-hand sides of the equations (6) for $0 \leqslant k \leqslant n$. The (possibly new) polynomials p' and q' so defined are in $A[z]$. Let F be the field of quotients of A and d_0 the greatest common divisor of p' and q' relative to $F[z]$. If $p_0 = p'/d_0$ and $q_0 = q'/d_0$ then there exist u_0 and v_0 in $F[z]$ such that $p_0u_0 + q_0v_0 = 1$. If k_0 is a common multiple of the denominators of the coefficients appearing in the polynomials u_0, p_0, v_0, and q_0, then $(k_0p_0)(k_0u_0) + (k_0q_0)(k_0v_0) = k_0^2$. Set $k = k_0^2$, $p = k_0p_0$, $q = k_0q_0$, $u = k_0u_0$, and $v = k_0v_0$. Then $f = p/q$ and

$$pu + qv = k \tag{7}$$

where p, u, q, and v are in $A[z]$ and $k \in A$. We can assume that the coefficients of q have no nontrivial common divisor since if they have one, say d, the equations of the type in (6) for $0 \leqslant k \leqslant n$ show that d also divides the coefficients of p and so could be cancelled from both p and q without affecting the equation $f = p/q$. Furthermore, equation (7) remains valid but with a possibly different element of A in place of k, if we divide both p and q by d.

From (7) we have $k = q(fu + v)$. Now it is well known that if k is an irreducible divisor of all the coefficients of the product of two power series with coefficients in a unique factorization domain, then it divides all the coefficients of at least one of the series. Thus if e is the constant term of the power series representing $(fu + v)$ we have $k|e$ as well as $k = b_0e$. This shows that b_0 is a unit of A. $\square$

This result naturally brings up the question of just what are the discrete subrings of **C** with rank 2 and unique factorization. By Proposition 1.10 we know that a discrete subring of **C** with rank 2 is contained in the ring of integers of a unique imaginary quadratic field. The question of which rings of integers of imaginary quadratic fields have unique factorization was settled in Stark [**67**]. It is well known that the Gaussian integers $\mathbf{Z} + i\mathbf{Z}$ is one such ring.

CHAPTER 8

MUNTZ'S THEOREM AND INTEGRAL POLYNOMIALS

Müntz's well-known theorem can be stated as follows. Let $\Lambda = \{\lambda_i\}$ be a sequence of real numbers satisfying $0 < \lambda_1 < \lambda_2 < \dots$. By a Λ-polynomial we mean a function of the form

$$p(x) = a_0 + \sum_{i=1}^{n} a_i x^{\lambda_i} \tag{1}$$

where the a_i's are arbitrary real numbers.

THEOREM 8.1. *The Λ-polynomials are dense in $C[0, 1]$, i.e., every f in $C[0, 1]$ can be uniformly approximated by them, if and only if $\sum_{j=1}^{\infty} \lambda_j^{-1} = \infty$.*

For the sake of completeness we sketch a proof. First suppose that $\sum_{k=1}^{\infty} \lambda_k^{-1} = \infty$. In order to prove density it suffices, by the Hahn-Banach theorem, to show that for any bounded linear functional φ on $C[0, 1]$ with $\varphi(t^{\lambda_k}) = 0$, $1 \leqslant k$, we have $\varphi \equiv 0$. By the Riesz representation theorem (cf. Rudin **[74]**) we can represent φ in the form $\varphi(h) = \int h\, d\mu$, $h \in C[0, 1]$, where μ is a finite regular Borel measure on $[0, 1]$. We define $f(z) = \int t^z\, d\mu$. It is easily seen that f is a bounded holomorphic function in the right half-plane H. Also $f(\lambda_k) = \varphi(t^{\lambda_k}) = 0$, $1 \leqslant k$. Thus the functions

$$g_k(z) = f(z)\frac{\lambda_k + z}{\lambda_k - z}, \qquad 1 \leqslant k,$$

are holomorphic in H. We can suppose without loss of generality that $|f| \leqslant 1$. The second factor of g_k has modulus 1 on the imaginary axis and modulus at most $1 + \varepsilon$, for any preassigned $\varepsilon > 0$, on all sufficiently large circles centered at the origin. Thus the factor has modulus at most 1 throughout H and the same is true of g_k. Similarly the functions $f(z)\prod_{k=1}^{n}(\lambda_k + z)/(\lambda_k - z)$ have moduli bounded by 1 throughout H. But for $0 < x < \lambda_1$ we have

$$|f(x)| \leqslant \prod_{k=1}^{n} \frac{\lambda_k - x}{\lambda_k + x} \leqslant \exp\left(-2x \sum_{k=1}^{n} \frac{1}{\lambda_k}\right)$$

where the elementary inequality $\log x \leqslant x - 1$ is used in establishing the second inequality. Letting $n \to \infty$ we see that $f(x) = 0$. Since f is holomorphic, $f \equiv 0$

which shows in particular that $\varphi(t^n) = f(n) = 0$, $1 \leqslant n$. Since the ordinary polynomials are dense in $C[0, 1]$, this shows that $\varphi \equiv 0$ as was to be proved.

Now suppose that $\sum_{k=1}^{\infty} \lambda_k^{-1} < \infty$. Then by standard theorems on infinite products

$$f(z) = \frac{z}{(2+z)^3} \prod_{k=1}^{\infty} \frac{\lambda_k - z}{2 + \lambda_k + z}$$

is a meromorphic function in the whole plane with poles at -2 and $-2 - \lambda_k$, $1 \leqslant k$, and zeros at 0 and λ_k, $1 \leqslant k$. We will be done once we show that there is a finite regular Borel measure μ such that $f(z) = \int t^z \, d\mu(t)$ since we have that the closed linear span of $\{1, t^{\lambda_1}, t^{\lambda_2}, \ldots\}$ is in the null space of the corresponding bounded linear transformation $\varphi(h) = \int h \, d\mu$ and this null space is not all of $C[0, 1]$ (in fact it fails to contain each t^λ for which $\lambda \neq 0$ and $\lambda \neq \lambda_k$, $1 \leqslant k$). By an application of the Cauchy integral formula we can obtain

$$f(z) = -\frac{1}{2\pi} \int_{-\infty}^{\infty} \frac{f(-1 + is)}{-1 + is - z} \, ds, \qquad \operatorname{Re} z > -1.$$

(One first applies the formula to the right half-disc centered at -1 and of radius R and then lets $R \to \infty$.) If we then make the substitution $(1 + z - is)^{-1} = \int_0^1 t^{z - is} \, ds$, $\operatorname{Re} z > -1$, and apply Fubini's theorem, we obtain

$$f(z) = \int_0^1 t^z \left\{ \frac{1}{2\pi} \int_{-\infty}^{\infty} f(-1 + is) e^{-is(\log t)} \, ds \right\} dt.$$

The term in curly brackets is a bounded continuous function h on $(0, 1]$ (in fact it is a certain Fourier transform evaluated at $\log t$) and we have the desired representation of f upon taking $d\mu = h dt$.

DEFINITION 8.2. An *integral Λ-polynomial* is a function of the form (1) where all the a_i's are rational integers.

It is natural to ask if Müntz's theorem holds with *integral Λ-polynomials* in place of *Λ-polynomials*. The integral Λ-polynomials all take on integral values at both 0 and 1; hence any function approximated by them must have the same property. Let $C_0[0, 1]$ denote the continuous functions on $[0, 1]$ which take on integral values at both 0 and 1. The analogue of Müntz's theorem is the following.

THEOREM 8.3. *If the sequence Λ consists of integers, then the integral Λ-polynomials are dense in $C_0[0, 1]$ if and only if $\sum_{i=1}^{n} \lambda_i^{-1} = \infty$.*

It is immediate from Theorem 8.1 that the condition $\sum \lambda_i^{-1} = \infty$ is necessary for the density of the integral Λ-polynomials; hence we need prove only the converse. This will follow after a series of lemmas.

DEFINITION 8.4. If q and s are positive integers, $q < s$, then define

$$D_{qs} = \inf_{c_j} \left\| x^{\lambda_q} - \sum_{j=q+1}^{s} c_j x^{\lambda_j} \right\|_{[0,1]}$$

where the infimum is taken over all possible choices of the coefficients c_j, $q + 1 \leqslant j \leqslant s$.

We note that D_{qs} is the distance in the uniform metric from the function x^{λ_q} to the subspace spanned by $x^{\lambda_{q+1}}, \ldots, x^{\lambda_s}$.

LEMMA 8.5. *If q and s are positive integers, $q < s$, then*

$$D_{qs} \leqslant \exp\left(-2\lambda_q \sum_{j=q+1}^{s} \lambda_j^{-1} \right).$$

PROOF. There exists a Λ-polynomial p of the form $p(x) = \sum_{j=q+1}^{s} c_j x^{\lambda_j}$ such that ($\|\cdot\| = \|\cdot\|_{[0,1]}$)

$$\|x^{\lambda_q} - p(x)\| \leqslant \prod_{j=q+1}^{s} \frac{\lambda_j - \lambda_q}{\lambda_j + \lambda_q} = \prod_{j=q+1}^{s} \frac{1 - \lambda_q/\lambda_j}{1 + \lambda_q/\lambda_j}$$

$$\leqslant \prod_{j=q+1}^{s} e^{-2\lambda_q/\lambda_j} = \exp\left(-2\lambda_q \sum_{j=q+1}^{s} \lambda_j^{-1} \right).$$

The second inequality here follows from the inequality $(1 - x)/(1 + x) \leqslant e^{-2x}$, $x \geqslant 0$, which can be proved by elementary means.

The first inequality requires more effort (von Golitschek [70, Lemma 2]). For any distinct real numbers $p, r_1, r_2, \ldots, r_k$, all $> -\frac{1}{2}$, we set

$$d(x^p, \{r_i\}) = \inf_{c_1, \ldots, c_k} \left\{ \int_0^1 \left(t^p - \sum_{i=1}^{k} c_i t^{r_i} \right)^2 dt \right\}^{1/2}$$

and if $p, r_1, \ldots, r_k$ are distinct and nonnegative, then we set

$$E(x^p, \{r_i\}) = \inf_{c_1, \ldots, c_k} \left(\max_{0 \leqslant x \leqslant 1} \left| x^p - \sum_{i=1}^{k} c_i x^{r_i} \right| \right).$$

The first inequality will follow once we establish that

$$E(x^p, \{r_i\}) \leqslant \prod_{i=1}^{k} \frac{|p - r_i|}{p + r_i}.$$

Let M be any positive number. Let $p' = Mp$ and $r_i' = Mr_i$, $1 \leqslant i \leqslant k$. Then for any real coefficients b_i and $0 \leqslant x \leqslant 1$ we have

$$J(x) = \left| x^{p'+1/2} - \sum_{i=1}^{k} b_i x^{r_i'+1/2} \right|$$

$$= \left(p' + \tfrac{1}{2}\right) \left| \int_0^x \left(t^{p'-1/2} - \sum_{i=1}^{k} c_i t^{r_i'-1/2} \right) dt \right|$$

where the first equality defines J and the c_i's are properly chosen. Applying Schwarz' inequality we obtain

$$J(x) \leqslant \left(p' + \tfrac{1}{2}\right)\sqrt{x} \left\{ \int_0^1 \left(t^{p'-1/2} - \sum_{i=1}^{k} c_i t^{r_i'-1/2} \right)^2 dt \right\}^{1/2}.$$

By a proper choice of the coefficients b_i and c_i we can minimize the last estimate and obtain

$$J(x) \leqslant \left(p' + \tfrac{1}{2}\right)\sqrt{x}\; d\left(x^{p'-1/2}, \{r_i' - \tfrac{1}{2}\}\right).$$

From the classical proof of Müntz's theorem (cf. Cheney [66, §6.2]) we find that

$$d\left(x^{p'-1/2}, \{r_i' - \tfrac{1}{2}\}\right) \leqslant \frac{1}{\sqrt{2p'}} \prod_{i=1}^{k} \frac{|p' - r_i'|}{p' + r_i'}.$$

Combining the last two estimates we find that

$$J(x) \leqslant \sqrt{x}\, \frac{p' + \frac{1}{2}}{\sqrt{2p'}} \prod_{i=1}^{k} \frac{|p' - r_i'|}{p' + r_i'}, \qquad 0 \leqslant x \leqslant 1.$$

In view of this we have

$$E\left(x^{p'}, \{r_i'\}\right) \leqslant \max_{0 \leqslant x \leqslant 1} \left| x^{p'} - \sum_{i=1}^{k} b_i x^{r_i'} \right| \leqslant \frac{p' + \frac{1}{2}}{\sqrt{2p'}} \prod_{i=1}^{k} \frac{|p - r_i'|}{p' + r_i'}.$$

For arbitrary coefficients a_i it is clear that

$$\max_{0 \leqslant x \leqslant 1} \left| x^{p} - \sum_{i=1}^{k} a_i x^{r_i} \right| = \max_{0 \leqslant x \leqslant 1} \left| x^{p'} - \sum_{i=1}^{k} a_i x^{r_i'} \right|$$

hence

$$E\left(x^{p}, \{r_i\}\right) = E\left(x^{p'}, \{r_i'\}\right).$$

From this and the last estimate of $E(x^{p'}, \{r_i'\})$ with M chosen to be $1/(2p)$ we obtain

$$E\left(x^{p}, \{r_i\}\right) \leqslant \frac{p' + \frac{1}{2}}{\sqrt{2p'}} \prod_{i=1}^{k} \frac{|p' - r_i'|}{p' + r_i'} = \prod_{i=1}^{k} \frac{|p - r_i|}{p + r_i}$$

as was to be proved. □

COROLLARY 8.6. *If q and s are as above, then there is a Λ-polynomial Q_{qs} of the form*

$$Q_{qs}(x) = \sum_{j=q+1}^{s} c_{qsj} x^{\lambda_j}$$

such that $Q_{qs}(1) = 1$ and

$$\|x^{\lambda_q} - Q_{qs}(x)\| = 2D_{qs} = A_{qs} \leqslant 2 \exp\left(-2\lambda_q \sum_{j=q+1}^{s} \lambda_j^{-1}\right)$$

where the second equality serves to define A_{qs}.

PROOF. The subspace of $C[0, 1]$ spanned by $x^{\lambda_{q+1}}, \ldots, x^{\lambda_s}$ is obviously finite dimensional so there exists, by a standard argument, a linear combination p of these powers of x which satisfies $\|x^{\lambda_q} - p(x)\| = D_{qs}$. If we set $Q_{qs}(x) = p(x) + (1 - p(1))x^{\lambda_1}$ then

$$\begin{aligned}\|x^{\lambda_q} - Q_{qs}(x)\| &= \|x^{\lambda_q} - p(x) - (1 - p(1))x^{\lambda_1}\| \\ &\leqslant \|x^{\lambda_q} - p(x)\| + |1 - p(1)| \leqslant 2\|x^{\lambda_q} - p(x)\| \\ &= 2D_{qs} = A_{qs}. \quad \square\end{aligned}$$

LEMMA 8.7. *Let r and s be positive integers, $r < s$, and suppose that $|\Sigma_{j=k}^{s} d_j| \leqslant 1$, $r + 1 \leqslant k \leqslant s$, and $\Sigma_{j=r+1}^{s} d_j = 0$. Then setting*

$$p_{rs}(x) = \sum_{j=r+1}^{s} d_j x^{\lambda_j}$$

we have

$$\|p_{rs}\| \leqslant (\lambda_s - \lambda_r)/\lambda_r.$$

PROOF. Since, by hypothesis, $p_{rs}(1) = \Sigma_{j=r+1}^{s} d_j = 0$ we have

$$p_{rs}(x) = \sum_{\kappa=r+1}^{s} (x^{\lambda_\kappa} - x^{\lambda_{\kappa-1}})\left(\sum_{j=\kappa}^{s} d_j\right)$$

and for $0 \leqslant x \leqslant 1$,

$$\begin{aligned}|p_{rs}(x)| &\leqslant \sum_{\kappa=r+1}^{s} |x^{\lambda_\kappa} - x^{\lambda_{\kappa-1}}| = \sum_{\kappa=r+1}^{s} (x^{\lambda_{\kappa-1}} - x^{\lambda_\kappa}) \\ &= x^{\lambda_r} - x^{\lambda_s} \leqslant (\lambda_s - \lambda_r)/\lambda_r.\end{aligned}$$

The last inequality is perhaps most easily proved as follows. Suppose $0 < \alpha < \beta$. Fix x $(0 \leqslant x \leqslant 1)$ and regard x^α as a function of α. Then by the mean value theorem, for some choice of γ we have

$$\frac{x^\alpha - x^\beta}{\alpha - \beta} = x^\gamma \ln x \leqslant x^\alpha \ln x, \qquad \alpha < \gamma < \beta.$$

The right-hand side of this inequality is zero at both $x = 0$ and $x = 1$, hence has its maximum absolute value on $[0, 1]$ at a zero of its derivative. There is exactly one such, at $x = e^{-1/\alpha}$, which leads to $x^\alpha - x^\beta \leqslant (\beta - \alpha)/\alpha e < (\beta - \alpha)/\alpha$. $\square$

The method of proof is roughly first to select a Λ-polynomial of best approximation and then to replace the monomials which make it up by integral Λ-polynomials. The replacements are made by means of repeated applications of Corollary 8.6 and a final application of Corollary 8.7. The replacement process is carried out in accordance with the following lemma.

DEFINITION 8.8. For every positive integer s and $f \in C[0, 1]$ let

$$E_s(f) = \inf_{a_j \in \mathbf{R}} \left\| f(x) - \left(a_0 + \sum_{j=1}^{s} a_j x^{\lambda_j}\right)\right\|. \tag{2}$$

LEMMA 8.9. *Let r and s be positive integers with $r < s$ and $f(1) = 0$. Then there exist integers b_j, $1 \leqslant j \leqslant s$, such that*

$$\left\| f(x) - \sum_{j=1}^{s} b_j x^{\lambda_j} \right\| \leqslant 2E_s(f) + \sum_{q=1}^{r} A_{qs} + \frac{\lambda_s - \lambda_r}{\lambda_r}$$

where the A_{qs}'s were defined in Corollary 8.6.

PROOF. Since in (2) we are approximating f by the elements of a finite dimensional space ($=\operatorname{span}\{x^{\lambda_1}, \ldots, x^{\lambda_s}\}$), there exists a Λ-polynomial $\tilde{p}_s$ of the form (1) with $n = s$ such that $\|f - \tilde{p}_s\| = E_s(f)$. Setting $p_s(x) = \tilde{p}_s(x) - \tilde{p}_s(1)x^{\lambda_1} = \sum_{j=1}^{s} a_{j0} x^{\lambda_j}$ we have $(|\tilde{p}_s(1)| = |f(1) - \tilde{p}_s(1)| \leqslant E_s(f))$

$$\|f - p_s\| \leqslant 2E_s(f) \tag{3}$$

and

$$p_s(1) = \sum_{j=1}^{s} a_{j0} = 0. \tag{4}$$

We next define certain coefficients b_j and a_{jq} by induction on q. By (3) and (4) we have (5) and (6) below when $q = 0$:

$$\left\| f(x) - \sum_{j=1}^{q} b_j x^{\lambda_j} - \sum_{j=q+1}^{s} a_{jq} x^{\lambda_j} \right\| \leqslant 2E_s(f) + \sum_{j=1}^{q} \|x^{\lambda_j} - Q_{js}(x)\| = A_q \tag{5}$$

and

$$\sum_{j=1}^{q} b_j + \sum_{j=q+1}^{s} a_{jq} = 0 \tag{6}$$

where the equality in (5) serves to define A_q. To describe the induction step we assume that (5) and (6) hold. Define $b_{q+1} = [a_{q+1,q}]$ and

$$a_{j,q+1} = a_{jq} + (a_{q+1,q} - b_{q+1})c_{q+1,s,j}, \qquad q + 2 \leqslant j \leqslant s,$$

where $\{c_{q+1,s,j}\}$ are the coefficients of the polynomial $Q_{q+1,s}$ in Corollary 8.6. Then

$$\begin{aligned}
&\left\| f(x) - \sum_{j=1}^{q+1} b_j x^{\lambda_j} - \sum_{j=q+2}^{s} a_{j,q+1} x^{\lambda_j} \right\| \\
&\quad = \left\| f(x) - \sum_{j=1}^{q} b_j x - \sum_{j=q+1}^{s} a_{jq} x^{\lambda_j} - (a_{q+1,q} - b_{q+1})\left(Q_{q+1,s}(x) - x^{\lambda_{q+1}}\right) \right\| \\
&\quad \leqslant A_q + \|Q_{q+1,s}(x) - x^{\lambda_{q+1}}\| = A_{q+1}
\end{aligned}$$

and, using (6),

$$\begin{aligned}
\sum_{j=1}^{q+1} b_j + \sum_{j=q+2}^{s} a_{j,q+1} &= \sum_{j=1}^{q+1} b_j + \sum_{j=q+2}^{s} \left(a_{jq} + (a_{q+1,q} - b_{q+1})c_{q+1,s,j}\right) \\
&= b_{q+1} - a_{q+1,q} + (a_{q+1,q} - b_{q+1}) \sum_{j=q+2}^{s} c_{q+1,s,j} = 0
\end{aligned}$$

since $1 = Q_{q+1,s}(1) = \sum_{j=q+2}^{s} c_{q+1,s,j}$. Thus (5) and (6) hold for $q + 1$ in place of q for this definition of b_{q+1} and $a_{j,q+1}$ $(q + 2 \leqslant j \leqslant s)$.

We stop the above induction at $q = r$ and proceed differently to define $b_{r+1}, \ldots, b_s$. Thus we have

$$\left\| f(x) - \sum_{j=1}^{r} b_j x^{\lambda_j} - \sum_{j=r+1}^{s} a_{jr} x^{\lambda_j} \right\| \leqslant 2E_s(f) + \sum_{j=1}^{r} \left\| x^{\lambda_j} - Q_{js}(x) \right\| = A_r \quad (7)$$

and

$$\sum_{j=1}^{r} b_j + \sum_{j=r+1}^{s} a_{jr} = 0. \quad (8)$$

Next define recursively, for $j = s, s-1, \ldots, r+1$, $d_s = a_{sr} - [a_{sr}]$ and

$$d_j = \begin{Bmatrix} a_{jr} - [a_{jr}] & \text{if } \Sigma_{i=j+1}^{s} d_j \leqslant 0 \\ a_{jr} - [a_{jr}] - 1 & \text{if } \Sigma_{i=j+1}^{s} d_j > 0 \end{Bmatrix}, \qquad s - 1 \geqslant j \geqslant r + 1. \quad (9)$$

By definition of the d_j's, $\Sigma_{j=r+1}^{s} d_j \equiv \Sigma_{j=r+1}^{s} a_{jr}$ (mod 1) and by (8) $\Sigma_{j=r+1}^{s} a_{jr} \equiv 0$ (mod 1). Also by definition we have $|\Sigma_{j=r+1}^{s} d_j| < 1$; hence

$$\sum_{j=r+1}^{s} d_j = 0.$$

Define $b_j = a_{jr} - d_j$, $r + 1 \leqslant j \leqslant s$, and $p_{rs}(x) = \Sigma_{j=r+1}^{s} d_j x^{\lambda_j}$. Then the polynomial p_{rs} satisfies the hypotheses of Lemma 8.7; hence $\|p_{rs}\| \leqslant (\lambda_s - \lambda_r)/\lambda_r$. Thus

$$\left\| f(x) - \sum_{j=1}^{s} b_j x^{\lambda_j} \right\| = \left\| f(x) - \sum_{j=1}^{r} b_j x^{\lambda_j} - \sum_{j=r+1}^{s} a_{jr} x^{\lambda_j} + p_{rs}(x) \right\|$$
$$\leqslant A_r + \|p_{rs}\| \leqslant A_r + (\lambda_s - \lambda_r)/\lambda_r. \quad \square$$

It remains to show that the right-hand side of the inequality in Lemma 8.9 can be made arbitrarily small by a proper choice of r and s. To this end we prove the following technical result.

LEMMA 8.10. *Suppose that* Λ *is an increasing sequence of positive integers with* $\Sigma \lambda_j^{-1} = \infty$, $0 < \varepsilon < \frac{1}{25}$, $K > 0$ *and* N *is an integer satisfying* $N \geqslant 1 + 1/\varepsilon^2$ *and* $\lambda_{N+1} \leqslant (N+1)^{5/4}$. *Then there exist integers* r *and* s *with* $N < r < s$

$$\lambda_s \leqslant s^{5/4}, \quad (10)$$

$$\lambda_s \leqslant (1 + 4\varepsilon)\lambda_r, \quad (11)$$

$$\sum_{j=N}^{s} \lambda_j^{-1} \geqslant K, \quad (12)$$

and

$$\lambda_q \sum_{j=q+1}^{s} \lambda_j^{-1} > \frac{\varepsilon}{12}\sqrt{q}\,, \qquad N \leqslant q \leqslant r. \quad (13)$$

PROOF. From the condition $\Sigma \lambda_j^{-1} = \infty$ it follows that we do not have $\lambda_j \geqslant j^{5/4}$ for all sufficiently large j. We will use this fact several times without explicit mention of it. Thus we can choose an integer M such that

$$M > N, \qquad |\lambda_M| \leqslant M^{5/4}, \quad \text{and} \quad \sum_{j=N}^{M} \lambda_j^{-1} \geqslant K + 2. \tag{14}$$

Claim 1. There exists an integer s_0, $N < s_0 \leqslant M$, satisfying

$$\lambda_{s_0} \leqslant s_0^{5/4}, \tag{15}$$

$$\sum_{j=N}^{s_0} \lambda_j^{-1} \geqslant K + 1, \tag{16}$$

and

$$\lambda_q \sum_{j=q+1}^{s_0} \lambda_j^{-1} + 5\lambda_q / s_0^{1/4} > \sqrt{q}\,, \qquad N \leqslant q < s_0. \tag{17}$$

To prove Claim 1 set

$$Q = \left\{ q | N \leqslant q < M, \lambda_q \sum_{j=q+1}^{M} \lambda_j^{-1} \leqslant \sqrt{q} \right\}. \tag{18}$$

If Q is empty take $s_0 = M$. Suppose not. Define $M^* = \min Q$.
Then from (18),

$$\sum_{j=M^*+1}^{M} \lambda_j^{-1} \leqslant \frac{\sqrt{M^*}}{\lambda_{M^*}} \leqslant \frac{1}{\sqrt{M^*}}. \tag{19}$$

Define $s_0 = \max\{q | N \leqslant q \leqslant M^*, \lambda_q \leqslant q^{5/4}\}$. By hypothesis this set is not empty and $N < s_0 \leqslant M^* < M$. Since $\lambda_q > q^{5/4}$ whenever $s_0 + 1 \leqslant q \leqslant M^*$, we have

$$\sum_{j=s_0+1}^{M^*} \lambda_j^{-1} < \int_{s_0}^{\infty} x^{-5/4}\, dx = 4 s_0^{-1/4}. \tag{20}$$

From (11) and (12) we have

$$\sum_{j=s_0+1}^{M} \lambda_j^{-1} < \frac{4}{s_0^{1/4}} + \frac{1}{\sqrt{M^*}} < \frac{5}{s_0^{1/4}} \leqslant 1. \tag{21}$$

This together with (14) establishes (16). Inequality (15) follows from the definition of s_0. From the definition of M^* and $s_0 \leqslant M^*$ we have

$$\lambda_q \sum_{j=q+1}^{M} \lambda_j^{-1} > \sqrt{q}\,, \qquad N \leqslant q < s_0.$$

Inequality (17) follows from this and (21) which completes the proof of Claim 1.
We next define, by induction, a finite sequence $s_1, s_2, \ldots, s_{\kappa+1}$ satisfying

$$s_{j+1} + [\varepsilon s_{j+1}] = s_j \text{ or } s_j - 1, \qquad 0 \leqslant j \leqslant \kappa, \tag{22}$$

and $s_{\kappa+1} \leqslant N < s_\kappa$.

Since $s_j > N > 1 + 1/\varepsilon^2$ and $\varepsilon < \frac{1}{25}$ by hypothesis, the sequence $\{s_j\}_{j=0}^{\kappa+1}$ is strictly increasing. It is also well defined since the left-hand side of (22) decreases by at most 2 when s_{j+1} is decreased by 1.

Claim 2. Let $1 \leqslant k \leqslant \kappa$. If

$$\lambda_{s_j} > (1 + 4\varepsilon)\lambda_{s_{j+1}}, \qquad 0 \leqslant j \leqslant k - 1, \tag{23}$$

then

$$\lambda_{s_j} \leqslant s_j^{5/4}, \qquad 0 \leqslant j \leqslant k, \tag{24}$$

and

$$\sum_{j=s_k+1}^{s_0} \lambda_j^{-1} \leqslant \frac{1}{2}\frac{s_k}{\lambda_{s_k}}. \tag{25}$$

To prove Claim 2 first notice that inequality (24) holds for $k = 0$ by Claim 1. We proceed by induction and assume the claim is true as stated. By the induction hypothesis and (23)

$$\lambda_{s_{k+1}} < \frac{\lambda_{s_k}}{1 + 4\varepsilon} < \frac{s_k^{5/4}}{1 + 4\varepsilon}.$$

Hence by (14)

$$\lambda_{s_{k+1}} \leqslant \frac{(1 + \varepsilon + 1/s_{k+1})^{5/4}}{1 + 4\varepsilon} s_{k+1}^{5/4} \leqslant s_{k+1}^{5/4}$$

where the second inequality can be verified by taking logarithms and noting that $(x - 1) \geqslant \ln x \geqslant (x - 1)/2$ for $1 \leqslant x \leqslant 2$.

Inequality (25) is established as follows. Using (14) we see that

$$\sum_{i=s_j+1}^{s_{j-1}} \lambda_i^{-1} \leqslant (s_{j-1} - s_j)\lambda_{s_j}^{-1} \leqslant \left(\varepsilon + \frac{1}{s_j}\right)\frac{s_j}{\lambda_{s_j}}, \qquad 1 \leqslant j \leqslant \kappa + 1. \tag{26}$$

Also from (22)

$$s_j \leqslant \left(1 + \varepsilon + \frac{1}{s_{j+1}}\right)s_{j+1} \leqslant \left(1 + \varepsilon + \frac{1}{s_k}\right)s_{j+1}, \qquad 0 \leqslant j \leqslant k - 1;$$

hence

$$s_j \leqslant (1 + \varepsilon + 1/s_k)^{k-j}s_k, \qquad 0 \leqslant j \leqslant k. \tag{27}$$

Iterating on (23) gives $\lambda_{s_j} \geqslant (1 + 4\varepsilon)^{k-j}\lambda_{s_k}$, $0 \leqslant j \leqslant k$. This, together with (27) gives

$$\frac{s_j}{\lambda_{s_j}} \leqslant \left(\frac{1 + \varepsilon + 1/s_k}{1 + 4\varepsilon}\right)^{k-j}\frac{s_k}{\lambda_{s_k}}, \qquad 0 \leqslant j \leqslant k,$$

and by (26) we have

$$\sum_{i=s_j+1}^{s_{j-1}} \lambda_i^{-1} \leqslant (\varepsilon + 1/s_j)\left(\frac{1 + \varepsilon + 1/s_k}{1 + 4\varepsilon}\right)^{k-j}\frac{s_k}{\lambda_{s_k}}, \qquad 1 \leqslant j \leqslant k.$$

Thus

$$\sum_{i=s_k+1}^{s_0} \lambda_i^{-1} = \sum_{j=k}^{1} \sum_{i=s_j+1}^{s_{j-1}} \lambda_i^{-1} \leqslant \sum_{j=k}^{1} (\varepsilon + 1/s_k)\left(\frac{1+\varepsilon+1/s_k}{1+4\varepsilon}\right)^{k-j} \frac{s_k}{\lambda_{s_k}}$$

$$\leqslant \frac{s_k}{\lambda_{s_k}}(\varepsilon + 1/s_k) \sum_{j=0}^{k-1} \left(\frac{1+\varepsilon+1/s_k}{1+4\varepsilon}\right)^j.$$

But $s_k > N > 1/\varepsilon^2$, so

$$\sum_{i=s_k+1}^{s_0} \lambda_i^{-1} \leqslant \frac{s_k}{\lambda_{s_k}}(\varepsilon+\varepsilon^2)\left(1 - \left(\frac{1+\varepsilon+\varepsilon^2}{1+4\varepsilon}\right)\right)^{-1} = \frac{s_k}{\lambda_{s_k}}\varepsilon(1+\varepsilon)\frac{1+4\varepsilon}{\varepsilon(3-\varepsilon)}$$

$$\leqslant \frac{s_k}{\lambda_{s_k}}(1+1/25)\frac{1+4/25}{3-1/25} < \frac{1}{2}\frac{s_k}{\lambda_{s_k}}$$

which establishes (25), hence Claim 2.

We have, using (22),

$$\sum_{i=N}^{s_\kappa} \lambda_i^{-1} \leqslant (1 + s_\kappa - s_{\kappa+1})\lambda_N^{-1} \leqslant \frac{2+\varepsilon s_{\kappa+1}}{N} \leqslant \frac{2+\varepsilon N}{N} = \frac{2}{N} + \varepsilon < 2\varepsilon.$$

This, together with (16), shows that (25) does not hold with $k = \kappa$. Thus, by Claim 2, (23) does not hold for $k = \kappa$ and we can define l to be the smallest integer satisfying $0 \leqslant l < \kappa$ and

$$\lambda_{s_l} \leqslant (1+4\varepsilon)\lambda_{s_{l+1}}. \tag{28}$$

Setting $s = s_l$ and $r = s_{l+1}$ we see that (10) through (13) are satisfied as follows.

If $l = 0$ then $\lambda_s \leqslant s^{5/4}$ by (15), $\lambda_s \leqslant (1+4\varepsilon)\lambda_r$ by (28), and by (16) we have (12). Otherwise $l \geqslant 1$ and (23) is valid for $0 \leqslant j \leqslant l-1$. From (24) it follows that $\lambda_{s_l} = \lambda_s \leqslant s_l^{5/4} = s^{5/4}$. Also, $\lambda_s \leqslant (1+4\varepsilon)\lambda_r$ follows from (28) and, from (16) and (25), we have

$$\sum_{i=N}^{s} \lambda_i^{-1} = \sum_{i=N}^{s_0} \lambda_i^{-1} - \sum_{i=s+1}^{s_0} \lambda_i^{-1} \geqslant K + 1 - \frac{1}{2}\frac{s_l}{\lambda_{s_l}} > K$$

which establishes (12).

It remains only to verify (13). We have that $r + [\varepsilon r] = s$ or $s - 1$; hence $[\varepsilon r] \leqslant s - r$ and it follows that

$$s - r \geqslant \varepsilon s/2. \tag{29}$$

Thus, for $N \leqslant q \leqslant r$,

$$\sum_{j=q+1}^{s} \lambda_j^{-1} \geqslant \sum_{j=r+1}^{s} \lambda_j^{-1} \geqslant \frac{s-r}{\lambda_s} \geqslant \frac{\varepsilon s}{2\lambda_s}.$$

Now from (17) and (25), since $s_0^{-1/4} \leqslant s^{-1/4} \leqslant s/\lambda_s$,

$$\sqrt{q} < \lambda_q \sum_{j=q+1}^{s_0} \lambda_j^{-1} + 5\lambda_q s_0^{-1/4} \leqslant \lambda_q \left\{ \sum_{j=q+1}^{s} \lambda_j^{-1} + \sum_{s+1}^{s_0} \lambda_j^{-1} + \frac{5s}{\lambda_s} \right\}$$

$$\leqslant \lambda_q \left\{ \sum_{j=q+1}^{s} \lambda_j^{-1} + \frac{s}{2\lambda_s} + \frac{5s}{\lambda_s} \right\}.$$

Hence using (29) we obtain

$$\sqrt{q} < \lambda_q \left\{ \sum_{j=q+1}^{s} \lambda_j^{-1} \right\} \left(1 + \frac{11}{2} \cdot \frac{2}{\varepsilon} \right).$$

It follows that for $N \leqslant q \leqslant r$,

$$\lambda_q \sum_{j=q+1}^{s} \lambda_j^{-1} > \sqrt{q} \left(1 + \frac{11}{\varepsilon} \right)^{-1} > \sqrt{q}\, \varepsilon/12$$

which establishes (13) and concludes the proof of the lemma. □

PROOF OF THEOREM 8.3 (CONCLUSION). Let $f \in C_0[0, 1]$. Since it suffices to approximate $f - (f(1)x^{\lambda_1} + f(0)(1 - x^{\lambda_1}))$, we can assume without loss of generality that $f(0) = 0 = f(1)$. Let $0 < \varepsilon < \frac{1}{25}$. By Theorem 8.1 we have that $E_s(f) \to 0$ as $s \to \infty$. Also $\lambda_j < j^{5/4}$ for infinitely many j or else we would have $\sum_{j=1}^{\infty} \lambda_j^{-1} < \infty$. Thus there exists an integer N such that

$$E_N(f) \leqslant \varepsilon, \quad N \geqslant 4!\,(6/\varepsilon)^5, \quad \text{and} \quad \lambda_N < N^{5/4}. \tag{30}$$

Choose $K > 0$ such that $\exp(-2K) \leqslant \varepsilon$. By Lemma 8.10 there exist integers r and s such that $N < r < s$ and (10) through (13) hold. Applying Lemma 8.9 with these integers r and s we see that there exist integers b_j $(1 \leqslant j \leqslant s)$ such that the conclusion of that lemma holds. We estimate the right-hand side of the inequality there as follows.

From (30) we have

$$2E_s(f) \leqslant 2E_n(f) \leqslant 2\varepsilon, \tag{31}$$

and from (11):

$$(\lambda_s - \lambda_r)/\lambda_r \leqslant 4\varepsilon. \tag{32}$$

Using Corollary 8.6 and then (12)

$$\sum_{q=1}^{N-1} A_{qs} \leqslant 2 \sum_{q=1}^{N-1} \exp\left(-2\lambda_q \sum_{j=q+1}^{s} \lambda_j^{-1} \right)$$

$$\leqslant 2 \sum_{q=1}^{N-1} \exp(-2\lambda_q K) \leqslant 2 \sum_{q=1}^{N-1} \varepsilon^{\lambda_q} < 3\varepsilon. \tag{33}$$

Using Corollary 8.6 and then (13),

$$\sum_{q=N}^{r} A_{qs} \leqslant 2 \sum_{q=N}^{r} \exp\left(-2\lambda_q \sum_{j=q+1}^{s} \lambda_j^{-1} \right) \leqslant 2 \sum_{q=N}^{r} e^{-\varepsilon\sqrt{q}/6}$$

$$< 2 \sum_{q=N}^{r} 4! \left(\frac{6}{\varepsilon} \right)^4 \frac{1}{q^2} < 4 \cdot 4! \left(\frac{6}{\varepsilon} \right)^4 \frac{1}{N} < \varepsilon. \tag{34}$$

Thus Lemma 8.9 tells us that

$$\left\| f(x) - \sum_{j=1}^{s} b_j x^{\lambda_j} \right\| \leqslant 2\varepsilon + (3\varepsilon + \varepsilon) + 4\varepsilon = 10\varepsilon. \quad \square$$

We can prove results similar to Theorem 8.3 but for more general exponents, and much more simply, as follows.

For the remainder of the chapter, let Λ be any subset of the positive real numbers and define Λ-polynomials and integral Λ-polynomials as before.

THEOREM 8.11. *If the set Λ has a limit point x_0 with $0 < x_0 < \infty$, then the integral Λ-polynomials are dense in $C_0[0, 1]$.*

PROOF. Let $f \in C_0[0, 1]$ and $\varepsilon > 0$. As in the last proof we can assume without loss of generality that $f(0) = f(1) = 0$. Since x_0 is a positive limit point of Λ, it is evident that we can extract a sequence $\{\lambda_j\}$ from Λ which satisfies

$$\lambda_j \to x_0, \tag{35}$$

$$\{\lambda_j\} \text{ is monotone,} \tag{36}$$

$$\lambda_j > 1, \quad \text{all } j \quad \text{or} \quad \lambda_j < 1, \quad \text{all } j, \tag{37}$$

and

$$(\lambda_1 - \lambda_j)/\lambda_k < \varepsilon, \quad \text{all } j, k. \tag{38}$$

Since $x_0 > 0$, $\sum_{j=1}^{\infty} \lambda_j/(1 + \lambda_j^2) = \infty$. But according to Szász's form of Müntz's theorem if Λ is a set of complex numbers with positive real parts, then the Λ-polynomials are dense in $C[0, 1]$ if $\sum_{n=1}^{\infty} \operatorname{Re} \lambda_n/(1 + |\lambda_n|^2) = \infty$ and not dense in $C[0, 1]$ if $\sum_{n=1}^{\infty} (1 + \operatorname{Re} \lambda_n)/(1 + |\lambda_n|^2) < \infty$. See Szász [**16**] or Paley and Wiener [**34**, Theorem XV]. Thus there is a Λ-polynomial $p_0(x) = a + \sum_{j=1}^{n} b_j x^{\lambda_j}$ with

$$\|f - p_0\| < \varepsilon \tag{39}$$

and a constant. By (35), since $f(0) = 0$, $|a| < \varepsilon$; hence

$$\|p_0 - p_1\| < \varepsilon \tag{40}$$

where $p_1 = p_0 - a$. It is easy to see that we can write p_1 in the form

$$p_1(x) = cx^{\lambda_1} + \sum_{j=2}^{n} a_j (x^{\lambda_j} - x^{\lambda_{j-1}}).$$

From (39), (40), and the fact that $f(1) = 0$ we have $|c| < 2\varepsilon$ and

$$\|p_1 - p_2\| < 2\varepsilon \tag{41}$$

where $p_2 = p_1 - cx^{\lambda_1}$. Define an integral Λ-polynomial $[p_2]$ by

$$[p_2](x) = \sum_{j=2}^{n} [a_j](x^{\lambda_j} - x^{\lambda_{j-1}})$$

where $[a_j]$ denotes the greatest integer less than or equal to a_j. Then, with $(a_j) = a_j - [a_j]$,

$$|p_2(x) - [p_2](x)| = \left|\sum_{j=2}^{n} (a_j)(x^{\lambda_j} - x^{\lambda_{j-1}})\right| \leq \sum_{j=2}^{n} (a_j)|x^{\lambda_j} - x^{\lambda_{j-1}}|$$

$$\leq \sum_{j=2}^{n} |x^{\lambda_j} - x^{\lambda_{j-1}}| = \left|\sum_{j=2}^{n} (x^{\lambda_j} - x^{\lambda_{j-1}})\right| = |x^{\lambda_n} - x^{\lambda_1}| \quad (42)$$

where the second equality results from the monotonicity of x^{λ_i} (uniform in x) as i increases. This monotonicity, in turn, follows from (36) and (37).

As in the proof of Lemma 8.7 we see that $|x^{\lambda_n} - x^{\lambda_1}| \leq |\lambda_1 - \lambda_n|/\min\{\lambda_1, \lambda_n\}$; hence by (42) and (38)

$$\|p_2 - [p_2]\| \leq \varepsilon. \quad (43)$$

From (39), (40), (41), and (43), $\|f - [p_2]\| < 5\varepsilon$. □

Another direction in which the above results can be extended is the following. Let $C_0[0, \alpha]$, $\alpha < 1$, denote the real valued continuous functions on the interval $[0, \alpha]$ with integral values at 0 and $\|\cdot\|$ the supremum norm on $C_0[0, \alpha]$.

THEOREM 8.12. *Let Λ be a subset of the positive real numbers with no finite limit point and $\sum_{\lambda \in \Lambda} \lambda^{-1} = \infty$. Then the integral Λ-polynomials are dense in $C_0[0, \alpha]$ for any $\alpha < 1$.*

PROOF. Let $f \in C_0[0, \alpha]$ and $\varepsilon > 0$. Since Λ has no finite limit points there are only finitely many λ's in any bounded interval and we can assume without loss of generality that $\alpha^{\lambda} < \varepsilon$, $\lambda \in \Lambda$. Next extract from Λ a sequence $\{\lambda_j\}$ which is monotone increasing and satisfies $\sum \lambda_j^{-1} = \infty$; hence

$$\sum_j \frac{\lambda_j}{1 + \lambda_j^2} = \infty.$$

Proceeding as in the proof of Theorem 8.11, we construct a Λ-polynomial p_1 satisfying

$$\|f - p_1\| < 2\varepsilon. \quad (44)$$

Then

$$\|p_1 - [p_1]\| \leq \|x^{\lambda_1}\| + \|x^{\lambda_n} - x^{\lambda_1}\| \leq 2\alpha^{\lambda_1} + \alpha^{\lambda_n} \leq 3\varepsilon.$$

This and (44) give $\|f - [p_1]\| < 5\varepsilon$. □

CHAPTER 9

A STONE-WEIERSTRASS TYPE THEOREM

If X is any compact Hausdorff space, it is possible to obtain a theorem which is very much like the now classical Stone-Weierstrass theorem. We will prove this theorem here and give several applications of it. Throughout this chapter the word integer will mean rational integer. Among these applications will be the results of Håstad **[58]** concerning the case where X is a compact subset of Euclidean n-space and the approximation is by polynomials in the n coordinate functions with coefficients in $\mathbf{Z}$.

Recall that $C(X)$ denotes the algebra of continuous complex valued functions on X with the uniform norm. A subset $\mathfrak{F}$ of $C(X, \mathbf{R})$ is said to be *point separating* if for any two distinct points x and y of X there exists f in $\mathfrak{F}$ with $f(x) \neq f(y)$. Let R be any discrete subring of the complex numbers $\mathbf{C}$. We let $R[\mathfrak{F}]$ denote the polynomials in elements of $\mathfrak{F}$ with coefficients in R. Thus an element q of $R[\mathfrak{F}]$ has the form

$$q = \sum_{j_1=0}^{r_1} \cdots \sum_{j_k=0}^{r_k} a_{j_1 \cdots j_k} f_1^{j_1} \cdots f_k^{j_k}$$

where the a's belong to R. It is clear that $R[\mathfrak{F}]$ is the smallest subring of $C(X)$ which contains $\mathfrak{F}$. We will characterize the elements of $C(X)$ which can be uniformly approximated by elements of $R[\mathfrak{F}]$. In particular, we will see in Corollary 9.3 that whenever there exists an element q of $\mathbf{Z}[f]$ with $q \neq 0$ on X and $\|q\| < 1$, then every element of $C(X, \mathbf{R})$ can be uniformly approximated by elements of $\mathbf{Z}[\mathfrak{F}]$; that is, the smallest subring of $C(X, \mathbf{R})$ containing $\mathfrak{F}$ is dense in $C(X, \mathbf{R})$. In this sense, then, the result is similar to the Stone-Weierstrass theorem.

Let $B(X)$ be the set of all q in $R[\mathfrak{F}]$ such that $\|q\| < 1$ and $J(X) = \{x \in X: q(x) = 0 \text{ all } q \in B(X)\}$. It is clear that $J(X) \neq X$ if and only if $B(X)$ contains an element which is not identically zero on X. At the other extreme, for $R = \mathbf{Z}$, we have the following result.

PROPOSITION 9.1. *With X and $\mathfrak{F}$ as above, the set $J(X)$ is empty if and only if there is an element q of $B(X)$ with $q > 0$ everywhere on X.*

PROOF. If there is such a q in $B(X)$ then it is immediate from the definition that $J(X) = \varnothing$. Conversely, suppose $J(X) = \varnothing$. For each x in X we can choose a $q_x \in B(X)$ such that $q_x(x) \neq 0$. Then $\{X \setminus Z_{q_x}\}_{x \in X}$ is an open cover of X and by compactness there exists a finite subcover, say that associated with $q_{x_1}, \ldots, q_{x_m}$. For a large enough positive even integer k the polynomial $q = q_{x_1}^k + \cdots + q_{x_m}^k$ will be an element of $B(X)$ with $q > 0$. □

We say that an element f of $C(X)$ is *R-approximable* if for every $\varepsilon > 0$ there is an element q of $R[\mathfrak{F}]$ such that $\|f - q\| < \varepsilon$. We say that f is *R-interpolable* on a subset S of X if there is an element q of $R[\mathfrak{F}]$ such that $f \equiv q$ on S. We say simply "approximable" or "interpolable" when $R = \mathbf{Z}$. The main theorem of this chapter is the following.

THEOREM 9.2 (HEWITT AND ZUCKERMAN **[59]**). *An element f of $C(X, \mathbf{R})$ is approximable if and only if it is interpolable on $J(X)$.*

PROOF. Suppose f is approximable. Then there exists a sequence q_n of elements of $\mathbf{Z}[\mathfrak{F}]$ which converges uniformly to f. Let N be a positive integer such that $\|f - q_n\| < \frac{1}{2}$ whenever $n \geqslant N$. Then, for $n_1, n_2 \geqslant N$, $\|q_{n_1} - q_{n_2}\| < 1$; hence

$$q_{n_1} \equiv q_{n_2} \quad \text{on } J(X).$$

It is clear from this that $q_n \equiv f$ on $J(X)$ whenever $n \geqslant N$.

Conversely, suppose that f is interpolated on $J(X)$ by q in $\mathbf{Z}[\mathfrak{F}]$. It clearly suffices to approximate the difference $f - q$ which is identically zero on $J(X)$. Thus we assume without loss of generality that $f \equiv 0$ on $J(X)$. Let I be the ideal generated by $B(X)$. It is easy to see that every element of I has the form $\sum_{\mu=1}^m g_\mu q_\mu$ where, for each index μ, $g_\mu \in C(X, \mathbf{R})$ and $q_\mu \in B(X)$. It is well known that the closure I^- of I in $C(X, \mathbf{R})$ consists of all elements of $C(X, \mathbf{R})$ which are zero on some closed subset of X (cf. Hewitt and Ross **[63**, Theorem C.30**]**), and it is easy to see that this set is $J(X)$. Thus $f \in I^-$ and for any $\varepsilon > 0$ there exists an element $\sum_{\mu=1}^m g_\mu q_\mu$ of I as above such that

$$\left\| \sum_{\mu=1}^{m} g_\mu q_\mu - f \right\| < \frac{\varepsilon}{3}. \tag{1}$$

By the Stone-Weierstrass theorem (cf. Dunford and Schwartz **[58**, IV.6.15**]**), the elements g_μ of $C(X, \mathbf{R})$ are contained in the closed subalgebra of $C(X, \mathbf{R})$ generated by $\mathfrak{F}$. Thus there is a finite subset $\{f_1, \ldots, f_k\}$ of $\mathfrak{F}$ and real numbers $\beta_{\mu,\lambda_1,\ldots,\lambda_k}$ such that for all μ

$$\left\| g_\mu - \sum_{\lambda_1=0}^{r_1} \cdots \sum_{\lambda_k=0}^{r_k} \beta_{\mu,\lambda_1,\ldots,\lambda_k} f_1^{\lambda_1} \cdots f_k^{\lambda_k} \right\| < \frac{\varepsilon}{3m}. \tag{2}$$

Combining (1) and (2) we have

$$\left\| \sum_{\mu=1}^{m} \sum_{\lambda_1=0}^{r_1} \cdots \sum_{\lambda_k=0}^{r_k} \beta_{\mu,\lambda_1,\ldots,\lambda_k} f_1^{\lambda_1} \cdots f_k^{\lambda_k} q_\mu - f \right\| < \frac{2\varepsilon}{3} \tag{3}$$

since $\|q_\mu\| < 1$ for each q. Let

$$A = m(r_1 + 1) \cdots (r_k + 1) \max\{1, \|f_1^{r_1}\|, \ldots, \|f_k^{r_k}\|\}.$$

There are polynomials $q_{\mu,\lambda_1,\ldots,\lambda_k}$ in $\mathbf{Z}[t]$ such that

$$|\beta_{\mu,\lambda_1,\ldots,\lambda_k} t - q_{\mu,\lambda_1,\ldots,\lambda_k}(t)| < \varepsilon/3A \tag{4}$$

for every t in $[0, \|q_\mu\|]$. This is an immediate consequence of Pál's result as proved in the introduction or Theorem 5.4 since here $J([0, \|q_\mu\|]) = \{0\}$. Substituting $q_\mu(x)$ for t in (4), multiplying by $f_1^{\lambda_1} \cdots f_k^{\lambda_k}$, and summing over the appropriate λ's and μ's gives

$$\left| \sum_{\mu=1}^{m} \sum_{\lambda_1=0}^{r_1} \cdots \sum_{\lambda_k=0}^{r_k} \beta_{\mu,\lambda_1,\ldots,\lambda_k} f_1^{\lambda_1}(x) \cdots f_k^{\lambda_k}(x) q_\mu(x) \right.$$
$$\left. - \sum_{\mu=1}^{m} \sum_{\lambda_1=0}^{r_1} \cdots \sum_{\lambda_k=0}^{r_k} f_1^{\lambda_1}(x) \cdots f_k^{\lambda_k}(x) q_{\mu,\lambda_1,\ldots,\lambda_k}(q_\mu(x)) \right| < \frac{\varepsilon}{3} \tag{5}$$

for all x in X. Combining (5) and (3) gives

$$\left\| \sum_{\mu=1}^{m} \sum_{\lambda_1=0}^{r_1} \cdots \sum_{\lambda_k=0}^{r_k} f_1^{\lambda_1} \cdots f_k^{\lambda_k} q_{\mu,\lambda_1,\ldots,\lambda_k} \circ q_\mu - f \right\| < \varepsilon$$

and we are done. □

In view of Proposition 9.1 we have the following.

COROLLARY 9.3. *Let X and $\mathfrak{F}$ be as above. Then $\mathbf{Z}[\mathfrak{F}]$ is dense in $C(X, \mathbf{R})$ if and only if there exists an element q of $\mathbf{Z}[\mathfrak{F}]$ with $\|q\| < 1$ and $q(x) > 0$, all $x \in X$.*

When X is a compact subset of Euclidean space $\mathbf{R}^k$ and $\mathfrak{F}$ is the set of coordinate functions $x_1, \ldots, x_k$, this is Theorem 3 of Håstad [**58**]. Similarly his Theorem 4 is our Theorem 9.2.

The result in Theorem 9.2 is too general to yield very precise information about approximability. We turn now to some special cases in which we can describe the set $J(X)$ and the class of approximable functions more closely.

THEOREM 9.4. *Let X and $\mathfrak{F}$ be as above. If $J(f(X)) \subset \mathbf{Z}$ for all f in $\mathfrak{F}$, then*

$$J(X) = \{x \in X: f(x) \in \mathbf{Z}, \text{ all } f \in \mathfrak{F}\}. \tag{1}$$

PROOF. Let $\tilde{J}$ denote the right-hand side of (1). If $x \in \tilde{J}$, then for every q in $B(X)$, $q(x)$ is an integer with absolute value less than unity; hence $q(x) = 0$. Thus, by definition, $x \in J$ and we have established that $\tilde{J} \subset J$. Suppose $x \notin \tilde{J}$. Then there exists f in $\mathfrak{F}$ with $f(x)$ a noninteger. In particular, $f(x) \notin J(f(X))$ by hypothesis. It is easy to see that we can find a continuous function φ on $f(X)$ such that

$$\varphi(f(x)) = f(x) - [f(x)] = (f(x)),$$
$$\varphi(J(X)) \subset \{0\}, \qquad \|\varphi\|_{f(X)} < 1.$$

Applying Theorem 9.2 with $f(X)$ in place of X and $\mathfrak{F}$ consisting simply of the identity function we see there exists a q in $\mathbf{Z}[t]$ such that

$$\|q - \varphi\|_{f(X)} < \min\{1 - \|\varphi\|, (f(x))\}.$$

Then $q \circ f$ is an element of $\mathbf{Z}[\mathfrak{F}]$ with $\|q \circ f\| < 1$ and $(q \circ f)(x) \neq 0$; hence $x \notin J$. Thus $J \subset \tilde{J}$. $\square$

When $J(f(X)) \subset \{0, 1\}$ for all f in $\mathfrak{F}$, our Theorem 9.2 can be sharpened considerably. We require the following preliminary result.

PROPOSITION 9.5. *Let X be as above and $\mathfrak{F}$ a separating family of continuous functions on X, all of which assume only the values* 0 *and* 1. *Then every integral valued continuous function on X is a polynomial in functions in $\mathfrak{F}$ with integral coefficients.*

PROOF. Since X is compact every integral valued continuous function f on X is finitely valued. Also, by continuity, $f^{-1}(m)$ is a both open and closed subset of X for any integer m. Thus f is a finite linear combination with integral coefficients of characteristic functions of open-and-closed subsets of X. It suffices to show that such characteristic functions lie in $\mathbf{Z}[\mathfrak{F}]$. Let A be any nonvoid open-and-closed subset of X. Let $y \in A$ and $x \in X \setminus A$ (if $X \setminus A \neq \varnothing$). Then by hypothesis there exists $f \in \mathfrak{F}$ with $f(x) \neq f(y)$. Let $q_x = f$ if $f(x) = 1$ and $q_x = 1 - f$ if $f(x) = 0$. Then $q_x(x) = 1$ and $q_x(y) = 0$. Let $C_x = \{\tilde{x} \in X: q_x(\tilde{x}) = 0\}$. The family of open sets $\{C_x\}_{x \in A}$ forms an open cover for the compact set A; hence there exists a finite subcover $C_{x_1}, \ldots, C_{x_m}$. Set $h_x = 1 - q_{x_1} \cdots q_{x_m}$. Then $h_x(x) = 0$ and $h_x(\tilde{x}) = 1$, all $\tilde{x} \in A$. Set $D_x = \{\tilde{x} \in X: h_x(\tilde{x}) = 0\}$. Then $\{D_x\}_{x \in X \setminus A}$ is an open cover of $X \setminus A$ by subsets of $X \setminus A$; $X \setminus A$ is compact; and there exists a finite subcover $D_{x_1}, \ldots, D_{x_k}$. Let $\varphi = h_{x_1} \cdots h_{x_k}$. It is clear that φ is the characteristic function of A and also that $\varphi \in \mathbf{Z}[\mathfrak{F}]$. $\square$

THEOREM 9.6. *Let X and $\mathfrak{F}$ be as above and $J(f(X)) \subset \{0, 1\}$ for all f in $\mathfrak{F}$. If φ is any function in $C(X, \mathbf{R})$, then φ is approximable on X if and only if it assumes integral values everywhere on $J(X)$.*

PROOF. From Theorem 9.4 we have that $J(X) = \{x \in X: f(x) \in \mathbf{Z}, \text{all } f \in \mathfrak{F}\}$. It is clear that every element of $\mathbf{Z}[\mathfrak{F}]$ is integral valued on $J(X)$. Thus φ is approximable only if it too assumes integral values on $J(X)$.

Conversely, notice first that if $J(X)$ is void, there is nothing to prove: By Theorem 9.2 every function in $C(X, \mathbf{R})$ is approximable on X. For nonvoid $J(X)$ we have that $J(X)$ is a compact subspace of X and that $\mathfrak{F}$ is a point separating family of functions on $J(X)$ and satisfies the hypotheses of Proposition 9.5 there. Thus if φ is continuous on X and integral valued on $J(X)$, then by Proposition 9.5 it is interpolable on $J(X)$ and by Theorem 9.2 it is approximable on X. $\square$

We next consider the case where X is the Cartesian product of compact subsets of $\mathbf{R}$, and $\mathfrak{F}$ consists of the projections onto the coordinates. We start with the finite dimensional case. The following theorem is a generalization of Hewitt and Zuckerman [**59**, Theorem 6.7] and Håstad [**58**, Theorem 5] in the sense that arbitrary compact subsets of R are considered in place of intervals. We recall that the transfinite diameter of an interval is one-fourth of its length (Example 2.15).

THEOREM 9.7. *Let n be a positive integer and $X_1, \ldots, X_n$ compact subsets of $\mathbf{R}$, each with transfinite diameter $d(X_i) < 1$. Let $\mathfrak{F} = \{\pi_1, \ldots, \pi_n\}$ where π_i ($1 \leqslant i \leqslant n$) is the function on $\prod_{i=1}^n X_i$ defined by $\pi_i(x_1, \ldots, x_n) = x_i$. Then*

$$J\left(\prod_{i=1}^{n} X_i\right) = \prod_{i=1}^{n} J(X_i)$$

where $J(X_i)$ is defined as in the real case (equivalently, the $\mathfrak{F}$ for the definition of $J(X_i)$ in this chapter consists merely of the identity function).

PROOF. Let $X = \prod_{i=1}^n X_i$ and let $x = (x_1, \ldots, x_n)$ stand for a generic element of X. We first need to know that for each i ($1 \leqslant i \leqslant n$) there is a polynomial $q^{(i)}$ in a single variable with integral coefficients (i.e., $q^{(i)} \in \mathbf{Z}[t]$) such that $\|q^{(i)}\|_{X_i} < 1$ and, for t in X_i, $q^{(i)}(t) = 0$ if and only if $t \in J(X_i)$. This is a consequence of Theorem 6.3, but in this case it can be proved much more simply as follows. If $B(X) = \{0\}$ (actually impossible here), then we can take $q^{(i)} \equiv 0$. Otherwise let $0 \not\equiv q_1 \in B(X)$. Then q_1 has at most finitely many zeros and for each of the elements $t_1, \ldots, t_k$ of $(Z_{q_1} \cap X_i) \setminus J(X_i)$, if such there be, let $q_2, \ldots, q_{k+1}$ be elements of $B(X)$ with $q_{i+1}(t_i) \neq 0$ ($1 \leqslant i \leqslant k$). Then for a large enough positive even integer m we can take $q^{(i)}$ to be the polynomial $q_1^m + \cdots + q_{k+1}^m$. Notice that $q^{(i)} \geqslant 0$ although we will not need this fact.

Since each $(q^{(i)} \circ \pi_i) \in B(X)$ we have

$$J(X) \subset \bigcap_{i=1}^{n} \{x \in X: (q^{(i)} \circ \pi_i)(x) = 0\} = \prod_{i=1}^{n} J(X_i). \tag{1}$$

If some $J(X_i)$ is void, then by (1) so is $J(X)$ and the theorem holds. We therefore suppose that all $J(X_i)$ are nonvoid.

To establish the reverse inclusion in (1) we shall prove that if $f \in C(X, \mathbf{R})$, and f is approximable on X, then f is interpolable on $\prod_{i=1}^n J(X_i)$. The reverse inclusion in (1) is then a consequence of Theorem 9.2 as follows. We suppose

$$y \in \left[\prod_{i=1}^{n} J(X_i)\right] \setminus J(X)$$

and derive a contradiction. Let γ be any transcendental real number. Clearly there exists a φ in $C(X, \mathbf{R})$ with $\varphi(J(X)) \subset \{0\}$ and $\varphi(y) = \gamma$. Then by Theorem 9.2, φ is approximable on X; hence, by what we are going to prove, it is interpolable on $\prod_{i=1}^n J(X_i)$. This is a contradiction since $y \in \prod_{i=1}^n J(X_i)$ and $q(y)$ is algebraic for all q in $\mathbf{Z}[\mathfrak{F}]$.

Thus let f be an element of $C(X, \mathbf{R})$ which is approximable on X. We will be done once we show that f is interpolable on $\prod_{i=1}^n J(X_i)$. For $1 \leqslant i \leqslant n$ let $\alpha_i = |\inf X_i|$, $\beta_i = |\sup X_i|$, and $J(X_i) = \{u_{i,1}, \ldots, u_{i,r_i}\}$. From Theorem 5.3 we see that the sets $J(X_i)$ are finite since $d(X_i) < 1$ by hypothesis. Let $\gamma = \max\{1, \alpha_1, \beta_1, \ldots, \alpha_n, \beta_n\}$ and let ε be any positive real number. Since f is approximable there exists a polynomial q_ε, with integral coefficients, in the functions $\pi_1, \ldots, \pi_n$ such that $\|f - q_\varepsilon\| < \varepsilon$. Thus for every point $(u_{1,j_1}, \ldots, u_{n,j_n})$ in $\Pi_{i=1}^n J(X_i)$ we have

$$|f(u_{1,j_1}, \ldots, u_{n,j_n}) - q_\varepsilon(u_{1,j_1}, \ldots, u_{n,j_n})| < \varepsilon. \tag{2}$$

Let $\lambda_1, \ldots, \lambda_n$ be any n nonnegative integers. If in (2) we multiply through by $u_{1,j_1}^{\lambda_1}, \ldots, u_{n,j_n}^{\lambda_n}$ and sum over all possible values of j_i we have

$$\Bigg| \sum_{j_1=1}^{r_1} \cdots \sum_{j_n=1}^{r_n} u_{1,j_1}^{\lambda_1} \cdots u_{n,j_n}^{\lambda_n} f(u_{1,j_1}, \ldots, u_{n,j_n})$$

$$- \sum_{j_1=1}^{r_1} \cdots \sum_{j_n=1}^{r_n} u_{1,j_1}^{\lambda_1} \cdots u_{n,j_n}^{\lambda_n} q_\varepsilon(u_{1,j_1}, \ldots, u_{n,j_n}) \Bigg| < \varepsilon\gamma^{\lambda_1 + \cdots + \lambda_n} r_1 \cdots r_n. \tag{3}$$

We claim that the second sum in (3) is an integer. Indeed, for each i $(1 \leqslant i \leqslant n)$, the set $\{u_{i,1}, \ldots, u_{i,r_i}\} = J(X_i) = J_0(X_i)$ by Theorem 5.3. The set $J_0(X_i)$ is a disjoint union of conjugate sets of algebraic integers over $\mathbf{Z}$ (Definition 3.4). The minimal polynomial of each conjugate set has integral coefficients; hence the entire set $J_0(X_i)$ is the set of roots of a monic polynomial with integral coefficients. Thus the elementary symmetric polynomials in $\{u_{i,1}, \ldots, u_{i,r_1}\}$ are integers. Notice that each sum

$$\sum_{j_n=1}^{r_n} u_{n,j_n}^{\lambda_n} q_\varepsilon(u_{1,j_1}, \ldots, u_{n,j_n}) \tag{4}$$

can be viewed as a symmetric polynomial in the variables $\{u_{n,1}, \ldots, u_{n,r_n}\}$ with coefficients in $\mathbf{Z}[u_{1,j_1}, \ldots, u_{n-1,j_{n-1}}]$. By the standard theorem on symmetric polynomials (Jacobson [**51**, Theorem 9, p. 109]) it can be written as a polynomial in the elementary symmetric polynomials of $\{u_{n,1}, \ldots, u_{n,r_n}\}$ with coefficients in $\mathbf{Z}[u_{1,j_1}, \ldots, u_{n-1,j_{n-1}}]$. Since, as we have seen, these elementary symmetric polynomials are integers, it is clear that the sum in (4) is of the form $q'(u_{1,j_1}, \ldots, u_{n-1,j_{n-1}})$ where q' is a polynomial with integral coefficients. Proceeding by induction we see that the entire second sum in (3) is an integer, as claimed. Next notice that the first sum in (3) is independent of ε, hence is a limit of a sequence of integers, hence an integer itself. If we take ε small enough to make the right-hand side of (3) less than unity, the two sums in (3) will be equal. Thus setting $\varepsilon = (2\gamma^{r_1 + \cdots + r_n - n} r_1 \cdots r_n)^{-1}$ forces

$$\sum_{j_1=1}^{r_1} u_{1,j_1}^{\lambda_1} \Bigg\{ \sum_{j_2=1}^{r_2} \cdots \sum_{j_n=1}^{r_n} u_{2,j_2}^{\lambda_2} \cdots u_{n,j_n}^{\lambda_n} \big(f(u_{1,j_1}, \ldots, u_{n,j_n})$$

$$- q_\varepsilon(u_{1,j_1}, \ldots, u_{n,j_n}) \big) \Bigg\} = 0 \tag{5}$$

for $0 \leqslant \lambda_i \leqslant r_i - 1$ $(1 \leqslant i \leqslant n)$. View (5) as a system of equations where the "unknowns" are the quantities within the braces and the coefficients are $u_{1,j_1}^{\lambda_1}$. The u_{1,j_1} $(1 \leqslant j_1 \leqslant r_1)$ are distinct by the way they were defined and $\det(u_{1,j_1}^{\lambda_1})$ $(1 \leqslant j_1 \leqslant r_1, 0 \leqslant \lambda_1 \leqslant r_1 - 1)$ is a Vandermonde determinant, hence nonzero. It follows that the quantities within the braces are all zero. Repeating this argument $n - 1$ times we have that

$$f(u_{1,j_1}, \ldots, u_{n,j_n}) = q_\varepsilon(u_{1,j_1}, \ldots, u_{n,j_n})$$

for all $(u_{1,j_1}, \ldots, u_{n,j_n})$ in $\prod_{i=1}^n J(X_i)$. $\square$

Theorem 9.7 can be strengthened in two ways. We will first relax the condition $d(X_i) < 1$, all i. Later we will also allow the factor spaces to be infinite in number. We need the following result which generalizes Håstad [**58**, Theorem 7] to infinite dimensions and which is interesting in its own right.

PROPOSITION 9.8. *Let* $\{X_\lambda\}_{\lambda \in \Lambda}$ *be a family of compact subsets of* $\mathbf{R}$ *with* $d(X_\lambda) \geqslant 1$, *all* $\lambda \in \Lambda$, *and set* $X = \prod_{\lambda \in \Lambda} X_\lambda$. *For each* μ *in* Λ *define* $\pi_\mu \colon X \to X_\mu$ *by* $\pi_\mu(\{x_\lambda\}_{\lambda \in \Lambda}) = x_\mu$ *and let* $\mathfrak{F} = \{\pi_\lambda\}_{\lambda \in \Lambda}$. *Then* $J(X) = X$; *hence an element of* $C(X, \mathbf{R})$ *is approximable if and only if it already lies in* $\mathbf{Z}[\mathfrak{F}]$.

PROOF. Suppose that $q \in B(X)$. Then q is a polynomial in only finitely many π_λ, say $\pi_{\lambda_1}, \ldots, \pi_{\lambda_n}$. We must prove that $q \equiv 0$. We proceed by induction on n. If $n = 1$ then q is an element of $\mathbf{Z}[x_{\lambda_1}]$ with $\|q\|_{X_{\lambda_1}} < 1$. As we saw in Proposition 4.14, since $d(X_{\lambda_1}) \geqslant 1$ we have that $q \equiv 0$ on X_{λ_1}. For $n > 1$ we assume that $q \not\equiv 0$ and derive a contradiction. Write $q(x) = q_m(\tilde{x})x_{\lambda_n}^m + q_{m-1}(\tilde{x})x_{\lambda_n}^{m-1} + \cdots + q_0(\tilde{x})$ where $\tilde{x}$ is a generic element of $\prod_{i=1}^{n-1} X_{\lambda_i}$, $q_i \in \mathbf{Z}[x_{\lambda_1}, \ldots, x_{\lambda_{n-1}}]$ $(0 \leqslant i \leqslant m)$, and x_{λ_n} is a generic element of X_{λ_n}. For a fixed $\tilde{x} \in \prod_{i=1}^{n-1} X_{\lambda_i}$, q is an element of $\mathbf{R}[x_{\lambda_n}]$ with uniform norm < 1; hence, its leading coefficient $q_m(\tilde{x})$ satisfies $|q_m(\tilde{x})| < 1$. Indeed, if not, then we could divide q by its leading coefficient and thereby obtain a monic polynomial on X_{λ_n} with uniform norm < 1, but this contradicts $d(X_{\lambda_n}) \geqslant 1$ by Theorem 2.11. Thus $\|q_m\|_{\prod_{i=1}^{n-1} X_{\lambda_i}} < 1$ and by the induction hypothesis $q_m \equiv 0$ which is a contradiction. $\square$

After the following theorem we will have established the conclusion of Theorem 9.7 under all possible values of the $d(X_i)$. (Notice that if $d(X_\lambda) \geqslant 1$, then $J(X_\lambda) = X_\lambda$ by Proposition 4.14.) The result is also a generalization of Håstad [**58**, Theorem 6]. Our proof is different from his.

THEOREM 9.9. *If in Theorem* 9.7 *we relax the hypothesis* $d(X_i) < 1$, *all* i, *to* $d(X_i) < 1$ *for at least one* i, *then the conclusion still holds.*

PROOF. Let $X = \prod_{i=1}^n X_i$ and $\tilde{x} \in X \setminus \prod_{i=1}^n J(X_i)$. Then there exists i_0 such that $\tilde{x}_{i_0} \notin J(X_{i_0})$. Also $d(X_{i_0}) < 1$ since $d(X_i) \geqslant 1$ implies $J(X_i) = X_i$ by Proposition 4.14. Without loss of generality we assume that $i_0 = 1$. Since $d(X_1) < 1$ there exists q in $\mathbf{Z}[t]$ with $\|q\|_{X_1} < 1$ and $J(X_1) = Z_q \cap X_1$, by Proposition 5.9. Then $(q \circ \pi_1) \in \mathbf{Z}[\mathfrak{F}]$, $\|q \circ \pi_1\|_X < 1$ and $(q \circ \pi_1)(\tilde{x}) \neq 0$ which shows that $\tilde{x} \notin J(X)$. Thus we have established $\prod_{i=1}^n J(X_i) \supset J(X)$. For the reverse inclusion let

$q \in \mathbf{Z}[\mathfrak{F}]$ with $\|q\|_X < 1$. Without loss of generality we assume that $d(X_i) < 1$, $1 \leqslant i \leqslant k$, and $d(X_i) \geqslant 1$, $(k+1) \leqslant i \leqslant n$. Then by Proposition 5.9, $J(X_i)$ is finite for $1 \leqslant i \leqslant k$ and we can write

$$J(X_i) = \{u_{i,1}, \ldots, u_{i,r_i}\}.$$

The elementary symmetric polynomials for the elements of $J(X_i)$ $(1 \leqslant i \leqslant k)$ are integers, as we saw in the proof of Theorem 9.7. Form the polynomial

$$\sum_{j_1=1}^{r_1} \cdots \sum_{j_k=1}^{r_k} q^{2l}(u_{1,j_1}, \ldots, u_{k,j_k}, x_{k+1}, \ldots, x_n). \tag{$*$}$$

For a sufficiently large positive integer l this polynomial will have uniform norm < 1 on X since $\|q\|_X < 1$. For each i $(1 \leqslant i \leqslant k)$ it is clearly a symmetric polynomial in $\{u_{i,1}, \ldots, u_{i,r_i}\}$; hence by the same induction argument as in the proof of Theorem 9.7, it is actually a polynomial $\tilde{q}$ in $\mathbf{Z}[x_{k+1}, \ldots, x_n]$. But

$$\|\tilde{q}\|_{\Pi_{i=k+1}^n X_i} < 1;$$

hence by Proposition 9.8, $\tilde{q} \equiv 0$. Since every summand in $(*)$ is nonnegative, this implies that

$$q(u_{1,j_1}, \ldots, u_{k,j_k}, x_{k+1}, \ldots, x_n) = 0$$

for every possible choice of $j_1, \ldots, j_k$ and $x_{k+1}, \ldots, x_n$. Thus

$$J(X) \supset \prod_{i=1}^{k} J(X_i) \times \prod_{i=k+1}^{n} X_i = \prod_{i=1}^{n} J(X_i)$$

where the last equality follows from the fact that $d(X_i) \geqslant 1$ implies $J(X_i) = X_i$ (Proposition 4.14). □

The following generalizes Hewitt and Zuckerman [59, Theorem 6.8] who proved it when the factor spaces are intervals of length < 4. Our proof is a modification of theirs.

THEOREM 9.10. *Let $\{X_\lambda\}_{\lambda \in \Lambda}$ be a family of compact subsets of* $\mathbf{R}$. *Define the coordinate projections $\{\pi_\lambda\}_{\lambda \in \Lambda}$ as in Proposition 9.8 and let $\mathfrak{F} = \{\pi_\lambda\}_{\lambda \in \Lambda}$. Then*

$$J\Big(\prod_{\lambda \in \Lambda} X_\lambda\Big) = \prod_{\lambda \in \Lambda} J(X_\lambda). \tag{1}$$

PROOF. Let $X = \Pi_{\lambda \in \Lambda} X_\lambda$ and $J_\lambda = J(X_\lambda)$, $\lambda \in \Lambda$. Suppose first that some J_λ is void, say for $\lambda = \tilde{\lambda}$. By Proposition 9.1 there is a q in $\mathbf{Z}[t]$ such that $\|q\|_{X_{\tilde{\lambda}}} < 1$ and $q > 0$ on $X_{\tilde{\lambda}}$. Then $(q \circ \pi_{\tilde{\lambda}}) \in \mathbf{Z}[\mathfrak{F}]$, $\|q \circ \pi_{\tilde{\lambda}}\|_X < 1$ and $(q \circ \pi_{\tilde{\lambda}}) > 0$ on X which shows that both sides of (1) are void, hence equal. Thus we can assume that every J_λ is nonvoid. Select an element u_λ from each J_λ.

Given a finite subset Φ of Λ, let L_Φ be the set of all x in X such that $x_\lambda = \mu_\lambda$ for $\lambda \notin \Phi$. Let $K_\Phi = L_\Phi \cap (\Pi_{\lambda \in \Lambda} J_\lambda)$.

Suppose that $f \in C(X, \mathbf{R})$ and f is approximable on X. Then for every $\varepsilon > 0$ there is a q_ε in $\mathbf{Z}[\mathfrak{F}]$ such that $\|q_\varepsilon - f\|_X < \varepsilon$. Let Λ_0 be the set of indices of the variables occurring in q_ε. Let Φ be any finite subset of Λ and $S = \Phi \cup \Lambda_0$. We define a function f' in the variables $\{x_\lambda\}_{\lambda \in S}$ as follows. For x' in $X_0 = \Pi_{\lambda \in S} X_\lambda$

let $f'(x') = f(x)$ where $x_\lambda = x'_\lambda$ if $\lambda \in \Phi$ and $x_\lambda = u_\lambda$ if $\lambda \notin \Phi$. Write $S = \{\lambda_1, \ldots, \lambda_n, \ldots, \lambda_k\}$ where $d(X_{\lambda_i}) < 1$ for $1 \leqslant i \leqslant n$ and $d(X_{\lambda_i}) \geqslant 1$ for $(n+1) \leqslant i \leqslant k$. As in the proof of Theorem 9.7, let $J(X_{\lambda_i}) = \{u_{i,1}, \ldots, u_{i,r_i}\}$ $(1 \leqslant i \leqslant n)$. Continuing as in that proof we have

$$\Big| \sum u_{1,j_1}^{\lambda_1} \cdots u_{n,j_n}^{\lambda_n} f'(u_{1,j_1}, \ldots, u_{n,j_n}, x_{n+1}, \ldots, x_k) - \sum u_{1,j_1}^{\lambda_1} \cdots u_{n,j_n}^{\lambda_n} q'_\varepsilon(u_{1,j_1}, \ldots, u_{n,j_n}, x_{n+1}, \ldots, x_k) \Big| < \varepsilon \gamma^{\lambda_1 + \cdots + \lambda_n} r_1 \cdots r_n \qquad (*)$$

where $x_{n+1}, \ldots, x_k$ are generic elements of $X_{\lambda_{n+1}}, \ldots, X_{\lambda_k}$, respectively, the λ_i's are any nonnegative integers, and the summations are over all possible j_i's, as before. By essentially the same induction argument about symmetric polynomials as before we see that the second sum in (*) is an element of $\mathbf{Z}[x_{n+1}, \ldots, x_k]$, say $q_\varepsilon^{\prime\wedge}$. For all sufficiently small ε, if q'_ε and $\tilde{q}'_\varepsilon$ satisfy (*), then the corresponding elements $q_\varepsilon^{\prime\wedge}$ and $\tilde{q}_\varepsilon^{\prime\wedge}$ of $\mathbf{Z}[x_{n+1}, \ldots, x_k]$ satisfy

$$\|q_\varepsilon^{\prime\wedge} - \tilde{q}_\varepsilon^{\prime\wedge}\| < 1,$$

where the uniform norm is taken over $\tilde{X} = \Pi_{(n+1) \leqslant i \leqslant k} X_{\lambda_i}$. By Proposition 9.8, $J(\tilde{X}) = \tilde{X}$; hence $q_\varepsilon^{\prime\wedge} - \tilde{q}_\varepsilon^{\prime\wedge} \equiv 0$ on $\tilde{X}$. Thus the first sum in (*) equals the second sum for any choice of $x_{n+1}, \ldots, x_k$ and sufficiently small ε. Continuing the argument as in the proof of Theorem 9.7, we finally conclude that

$$f' \equiv q'_\varepsilon \quad \text{on} \prod_{i=1}^{k} J(X_{\lambda_i}) \times \prod_{i=n+1}^{k} X_{\lambda_i};$$

that is $f' \equiv q'_\varepsilon$ on $\Pi_{\lambda \in S} J_\lambda$ since $X_{\lambda_i} = J_{\lambda_i}$ for $(n+1) \leqslant i \leqslant k$ by Proposition 4.14 and the fact that $d(X_{\lambda_i}) \geqslant 1$. Thus

$$f(x) = q_\varepsilon(x), \qquad x \in K_S,$$

and, in particular, f is matchable on K_Φ.

We claim that $K_\Phi \subset J(X)$. Indeed, if not, then there is a point y in $K_\Phi \setminus J(X)$. Let γ be any transcendental number. There is an element φ of $C(X, \mathbf{R})$ with $\varphi(y) = \gamma$ and $\varphi(J(X)) = \{0\}$ by Tietze's extension theorem. Such a φ is approximable by Theorem 9.2. By the above paragraph φ is matchable on K_Φ but this is a contradiction since $q(y)$ is an algebraic number for every q in $\mathbf{Z}[\mathcal{F}]$. This contradiction establishes the claim.

Thus we have $\Pi_{\lambda \in \Lambda} J_\lambda = (\cup K_\Phi)^- \subset J(X)$ where the union is taken over all finite subsets Φ of Λ. The reverse inclusion is easily established as in the proof of the previous theorem. □

We turn now to the case of approximation in L_p norms. A very general result obtains whenever the set $J(X)$ has zero measure, as it has in many interesting cases. For example (Proposition 3.9), it is finite whenever $d(X) < 1$. Notice that no arithmetic conditions are necessary here. Indeed, they would make no sense whenever singletons are null sets. However, a size condition remains, that is, the domain X can be too large. We saw in Theorem 2.12 that for an interval of length $\geqslant 4$ and Lebesgue measure, no nontrivial approximation is possible.

Suppose that X is a compact Hausdorff space and μ is a finite regular positive Borel measure on it. We denote by $L_p(\mu)$ $(1 \leqslant p < \infty)$ the space of real valued pth power integrable functions on X.

THEOREM 9.11. *We assume the same notation as in Theorem* 9.2. *If $J(X)$ is a μ-null set, then every f in $L_p(\mu)$ is approximable by elements of $Z[\mathfrak{F}]$.*

PROOF. First, it is a well-known consequence of Lusin's theorem that $C(X, \mathbf{R})$ is dense in $L_p(\mu)$; hence we can assume without loss of generality that f is continuous. By the (outer) regularity of μ, there exist open sets $V_1 \supset V_2 \supset \cdots \supset J$ such that $\mu(V_n) < 1/n$, all n; hence $\mu(\bigcap V_n) = 0$. For each n there is, by Urysohn's lemma, a ψ_n in $C(X, \mathbf{R})$ satisfying $0 \leqslant \psi_n \leqslant 1$, $\operatorname{supp} \psi_n \subset V_n$, and $\psi_n \equiv 1$ on J. Thus $\psi_n \to 0$ off $\bigcap V_n$, hence a.e. Setting $\varphi_n = 1 - \psi_n$ we have $\varphi_n \to 1$ a.e., hence $\varphi_n f \to f$ a.e. By Theorem 9.2, for each n, there exists a sequence $\{q_{nm}\}$ in $\mathbf{Z}[\mathfrak{F}]$ such that $q_{nm} \to \varphi_n f$ uniformly as $m \to \infty$. It follows that $q_{nn} \to f$ a.e. and that for sufficiently large n, $|q_{nn}| \leqslant |f| + 1$. By the Lebesgue dominated convergence theorem, $q_{nn} \to f$ in L_p norm. $\square$

CHAPTER 10

MISCELLANEOUS RESULTS

Some results do not seem to fit into any one of the preceding chapters. We recall that in the *general case* (Chapter 9) X is a fixed but arbitrary compact Hausdorff space, $\mathcal{F}$ is a point separating family in $C(X, \mathbf{R})$, and we are approximating by elements of $\mathbf{Z}[\mathcal{F}]$, the ring of polynomials in elements of $\mathcal{F}$ with rational integer coefficients. In the *complex case* X is a compact subset of the complex plane $\mathbf{C}$ and we are approximating by elements from $R[z]$ where R is a fixed but arbitrary discrete subring of $\mathbf{C}$ and z is the identity function. In the *real case* X is a compact subset of the real line $\mathbf{R}$ and we are approximating by elements from $\mathbf{Z}[x]$, x being the identity function. In all three cases the norm is $\|\cdot\| = \|\cdot\|_X$ where by definition $\|f\|_S = \sup_{x \in S}|f(x)|$, for any subset S of X. It is a convenient convention to take $\|f\|_\varnothing = 0$. It is also reasonable since the quantities $|f(x)|$ always lie in $[0, \infty)$ and if we take $[0, \infty)$ to be the universal set, then $\sup \varnothing = 0$ by definition. If n is any nonnegative integer, we denote the polynomials in elements of $\mathcal{F}$ with degree at most n and coefficients in a ring R by $(R[\mathcal{F}])_n$.

In all three cases we define a subset J of X by first setting $\mathcal{B} = \{q \in R[\mathcal{F}]: \|q\|_X < 1\}$ and then taking

$$J = \{x \in X: q(x) = 0, \text{all } q \in \mathcal{B}\}.$$

In general, for any $G \subset C(X), f \in C(X), S \subset X$, we define

$$\operatorname{dist}_S(f, G) = \inf_{g \in G} \|f - g\|_S.$$

We write $\operatorname{dist}(f, G)$ for $\operatorname{dist}_X(f, G)$. A best approximation to f from G is any $g^\circ \in G$ satisfying $\|f - g^\circ\| = \operatorname{dist}(f, G)$. The *existence question* is whether or not a best approximation exists for each given $f \in C(X)$ (or $C(X, \mathbf{R})$), and the *uniqueness question* is whether or not more than one best approximation exists. We will consider these questions and some others, as follows. In some cases it is possible to reduce the problem of determining the number $\operatorname{dist}(f, G)$ to that of inspecting a finite number of quantities. We will also show that in case it is not possible to approximate f uniformly, it may still be possible to do so pointwise at every point of $X \setminus J$.

Existence. As in the case of approximation by polynomials with arbitrary (say real) coefficients, the existence question has an affirmative answer if we approximate by polynomials of bounded degree. We start with the following. First note that a subset G of $C(X)$ or $C(X, \mathbf{R})$ has finite rank if the linear subspace it generates has finite dimension.

THEOREM 10.1. *If G is a closed subset of $C(X)$ of finite rank, then for every f in $C(X)$ there exists a best approximation in G.*

PROOF. Let $d = \operatorname{dist}(f, G)$. Then the set of best approximations to f from G is clearly $\bigcap_{d'>d}[(f + d'B) \cap G]$ where B is the closed unit ball of $C(X)$ ($B = \{f \in C(X): \|f\| \leqslant 1\}$). The problem is to show that this intersection is nonvoid. The family of sets involved clearly has the property that the intersection of any finite subfamily is nonvoid; hence it suffices to show that they are compact. Since $(f + d'B) \cap G$ is closed, it suffices to show that it is contained in a compact set. Since G has finite rank by hypothesis, the linear space V generated by it has finite dimension. Since V has finite dimension, it is locally compact. It follows that $(f + d'B) \cap V$ is compact and since it contains $(f + d'B) \cap G$ we are done. □

By imbedding $C(X, \mathbf{R})$ in $C(X)$ in the usual way, we immediately obtain the following.

COROLLARY 10.2. *If G is a closed subset of $C(X, \mathbf{R})$ of finite rank, then for every f in $C(X, \mathbf{R})$ there is a best approximation in G.*

These results apply in our three cases to give existence as follows.

In the complex case let $X \subset \mathbf{C}$ and consider the (complex) linear subspace spanned by $(R[z])_n$ where n is a nonnegative integer. Clearly the same space is spanned by the functions $1, z, \ldots, z^n$; hence $(R[z])_n$ has finite rank in $C(X)$. To be able to apply Theorem 10.1 with $G = (R[z])_n$, it suffices to show that $(R[z])_n$ is closed in $V = (\mathbf{C}[z])_n$ since the latter is finite dimensional, hence closed in $C(X)$. Suppose that $\{p_k\}$ is a sequence in $(R[z])_n$ converging to a polynomial p in V. Since the powers $1, z, \ldots, z^n$ form a basis for V the projections $\pi_i: V \to \mathbf{C}$ which send each polynomial into its ith coefficient exist. They are continuous on V since they are linear and V is finite dimensional. Thus $\pi_i(p_k) \to \pi_i(p)$ as $k \to \infty$. Since R is discrete and a subring of $\mathbf{C}$, it is closed in $\mathbf{C}$, as is well known. The $\pi_i(p_k)$ are elements of R; hence $\pi_i(p) \in R$, and this is for all i ($0 \leqslant i \leqslant n$). Thus $p \in (R[z])_n$. Since $\{p_k\}$ was any sequence in $(R[z])_n$ with limit in V (and V is metrizable, hence 1st countable, as are all finite dimensional topological vector spaces), $(R[z])_n$ is closed in V. Thus Theorem 10.1 applies and we have existence when approximating by elements of $(R[z])_n$.

In the real case the same argument applies, using Corollary 10.2 in place of Theorem 10.1; hence we have existence when approximating by elements of $(\mathbf{Z}[x])_n$ in $C(X, \mathbf{R})$ where n is any nonnegative integer.

In the general case, if we assume $\mathcal{F}$ to be finite, then $(\mathbf{Z}[\mathcal{F}])_n$ will have finite rank for any fixed but arbitrary nonnegative integer n and again the above method applies to show that $(\mathbf{Z}[\mathfrak{F}])_n$ is closed in $C(X, \mathbf{R})$; hence Corollary 10.2 applies.

Thus we have existence when approximating by integral coefficients and bounded degree.

Uniqueness. Approximation by polynomials with integral coefficients fails to be unique even in very simple cases. For example, consider approximation by $(Z[x])_n$ on a compact subset X of the real line with $0 \in X$. Let $f \equiv \frac{1}{2}$ on X and suppose n is any nonnegative integer. Then, since $0 \in X$, $\operatorname{dist}(f, (\mathbf{Z}[x])_n) \geqslant \frac{1}{2}$ and, since the identically zero function belongs to $(\mathbf{Z}[x])_n$, $\operatorname{dist}(f, (\mathbf{Z}[x])_n) = \frac{1}{2}$. The same argument shows that $\operatorname{dist}(f, \mathbf{Z}[x]) = \frac{1}{2}$. Thus, in all these cases the two polynomials $q = 0$ and $q = 1$ are best approximations to f on X. There may be even more. For example, if $n \geqslant 2$ in the above and $0 \in X \subset [-1, 1]$, then the polynomials $1 - x^2$ and x^2 are also best approximations.

The distance function. Here we will show that the function $\operatorname{dist}(f, G)$ can be simplified in many interesting cases. For what follows we must generalize the definition of J at the beginning of this chapter by substituting G for $R[\mathfrak{F}]$. We start with the following result.

THEOREM 10.3. *Suppose*

(i) *X is a compact Hausdorff space;*

(ii) *G is a subgroup of $C(X)$ (resp., $C(X, \mathbf{R})$);*

(iii) *an element f of $C(X)$ (resp., $C(X, \mathbf{R})$) is approximable* ($\operatorname{dist}(f, G) = 0$) *if and only if there exists q in G such that $f \equiv q$ on J.*

Then $f \in C(X)$ (resp., $C(X, \mathbf{R})$), $q \in G$, and $\|f - q\|_J < \|f - q\|_X$ imply that q is not a best approximation to f from G.

PROOF. Let δ be any positive number. Set $\mu = \|f - q\|_J$ and

$$F = \{x \in X: |f(x) - q(x)| \geqslant \mu + \delta\}.$$

By the continuity of f and q, hence $f - q$, F is a closed subset of X. Also notice from its definition that J is the intersection of the zeros of the elements of G with norm strictly less than one. Since these elements are continuous, J is also closed in X. From the definition of F we see that F and J are disjoint. Thus, by Urysohn's lemma there exists h in $C(X)$ with $0 \leqslant h \leqslant 1$, $h(J) \subset \{0\}$, and $h(F) \subset \{1\}$. Since $h(f - q) = 0$ on J there is by (iii) a $\tilde{q} \in G$ satisfying

$$\|h(f - q) - \tilde{q}\|_X < \delta. \tag{1}$$

Also, by the way h was constructed and the definition of F,

$$\begin{aligned}\|(f - q) - h(f - q)\|_X &= \|(f - q) - h(f - q)\|_{X \setminus F} \\ &\leqslant \|1 - h\|_{X \setminus F}\|f - q\|_{X \setminus F} \leqslant \mu + \delta. \end{aligned} \tag{2}$$

From (1), (2), and the triangle inequality $\|(f - q) - \tilde{q}\| < \mu + 2\delta$. Since δ was any positive number, we see that $\tilde{q}$ can be chosen so as to make $\|f - (q + \tilde{q})\|$ arbitrarily close to μ. Since $(q + \tilde{q}) \in G$ we are done. □

We see that Theorem 10.3 applies in all three of our cases, possibly under some further restrictions, as follows.

In the general case let G be the ring $\mathbf{Z}[\mathfrak{F}]$. We see from Theorem 9.2 that condition (iii) is satisfied; hence the conclusion of the theorem holds in this case.

In the complex case let A be a discrete subring of $\mathbf{C}$ of rank 2 and $G = A[z]$. If X is a subset of $\mathbf{C}$ with transfinite diameter $d(X) \geqslant 1$ then $\mathfrak{B} = \{0\}$ since, if $q \in A[z]$ and $0 < \|q\| < 1$ then dividing q by its leading coefficient would give a monic polynomial in $\mathbf{C}[z]$ with norm < 1, contradicting Theorem 2.11(i). Thus, if $d(X) \geqslant 1$ then $J = X$ and the conclusion of Theorem 10.3 holds vacuously. On the other hand, if $d(X) < 1$ and X is a Lavrent'ev subset (Definition 4.1) of $\mathbf{C}$, then by Theorem 4.7, condition (iii) of the present theorem is satisfied, hence the conclusion.

In the real case we see as in the complex case that if $d(X) \geqslant 1$ then the conclusion of Theorem 10.3 holds vacuously. If $d(X) < 1$ then by Theorem 5.4 condition (iii) of the present theorem is satisfied, hence its conclusion.

The following application of Theorem 10.3 is interesting in that it reduces the problem of determining the distance from an $f \in C(X)$ to the set of integral polynomials to a finite dimensional problem in many interesting cases. We first prove the following variation of Theorem 10.3.

THEOREM 10.4. *Under the hypotheses of Theorem* 10.3, *for any f in $C(X)$* (*respectively*, $C(X, \mathbf{R})$) *we have*

$$\operatorname{dist}_X(f, G) = \operatorname{dist}_J(f, G).$$

PROOF. It is clear from definitions that $\operatorname{dist}_X(f, G) \geqslant \operatorname{dist}_J(f, G)$. To prove the reverse inequality let ε be any positive number. Then by definition of $\operatorname{dist}_J(f, G)$ there exists $q \in G$ with $\|f - q\|_J < \operatorname{dist}_J(f, G) + \varepsilon/2$. If $\|f - q\|_X = \|f - q\|_J$ then by definition

$$\operatorname{dist}_X(f, G) < \operatorname{dist}_J(f, G) + \varepsilon/2. \tag{3}$$

If $\|f - g\|_X > \|f - g\|_J$ then as in the proof of Theorem 10.3 we can find $\tilde{q}$ in G such that

$$\|f - q - \tilde{q}\|_X \leqslant \|f - q\|_J + \varepsilon/2 < \operatorname{dist}_J(f, G) + \varepsilon.$$

Hence, since $(q + \tilde{q}) \in G$ we have

$$\operatorname{dist}_X(f, G) < \operatorname{dist}_J(f, G) + \varepsilon. \tag{4}$$

Since ε is any positive number, we conclude from either (3) or (4) that $\operatorname{dist}_X(f, G) \leqslant \operatorname{dist}_J(f, G)$ as was to be proved. □

COROLLARY 10.5. *Let X be a Lavrent'ev subset of* $\mathbf{C}$ *with $d(X) < 1$ and A a discrete subring of* $\mathbf{C}$ *with rank* 2 *and containing the identity. Then for any $f \in C(X)$*

(i) $\operatorname{dist}_X(f, A[z]) = \operatorname{dist}_J(f, A[z])$;

(ii) *there exists $q \in A[z]$ such that* $\|f - q\|_J = \operatorname{dist}_J(f, A[z])$.

PROOF. We have already seen that our hypotheses here imply those of Theorem 10.3, hence Theorem 10.4, and we have (i). By Proposition 3.9, J is finite. With this fact we can show (ii) as follows.

Certainly it suffices to show that in the infimum defining $\operatorname{dist}_J(f, G)$ ($G = A[z]$ here) we can replace G by a finite set. Let $q_0 \in G$. Then clearly

$$\inf_{q \in G} \|f - q\|_J = \inf_{\|f-q\| \leqslant \|f-q_0\|} \|f - q\|_J.$$

Also notice that $\|f - q\|_J = \|f - q|_J\|_J$ where $q|_J$ denotes the restriction of q to J. Thus it suffices to show that the set $S_1 = \{q|_J: \|f - q|_J\|_J \leqslant \|f - q_0\|_J\}$ is finite. But, for any $q|_J \in S_1$,

$$\|q|_J\|_J \leqslant \|f\|_J + \|f - q\|_J \leqslant \|f\|_J + \|f - q_0\|_J = M,$$

where the equality serves to define M. Thus S_1 is a subset of the set $S_2 = \{q|_J: \|q\| \leqslant M\}$ and it suffices to show that S_2 is finite. Clearly, $G|_J$ is a subset of $C(J)$. Since J is finite, $C(J)$ is finite dimensional; hence the set $S_3 = \{f \in C(J): \|f\|_J \leqslant M\}$ is compact. Also, $G|_J$, hence S_2, is a discrete subgroup of S_3 since $\|q_1|_J - q_2|_J\|_J < 1$ implies $q_1 \equiv q_2$ by definition of J. It follows that $G|_J$, hence S_2, is closed in $C(J)$. Thus S_2 is a closed discrete subset of the compact set S_3, hence finite. □

By the same method of proof we can establish the corresponding result in the real case. We state it for emphasis.

COROLLARY 10.6. *Let X be a compact subset of $\mathbf{R}$ with $d(X) < 1$. Then for any f in $C(X, \mathbf{R})$ we have*

(i) $\operatorname{dist}_X(f, \mathbf{Z}[x]) = \operatorname{dist}_J(f, \mathbf{Z}[x])$;

(ii) *there exists q in $\mathbf{Z}[x]$ such that* $\|f - q\|_J = \operatorname{dist}_J(f, \mathbf{Z}[x])$.

Actually, in the proof of Corollary 10.5 we have obtained more information than in the statement. Since this might be important in future applications of the theory, we carry the analysis further. Assume then that the hypotheses of Corollary 10.5 hold and also that $A = I_L$ for some imaginary quadratic field L.

First note that $M = M(f, q_0)$. We kept this general form for M since a judicious choice of q_0 might lead to a small value of M. It is always possible to choose $q_0 = 0$, however, which leads to $M = M(f, 0) = 2\|f\|_J$.

Next let $J = \{\zeta_1, \ldots, \zeta_n\}$ and let $\tilde{q}$ be the monic polynomial with the elements of J as simple roots. Then by Theorem 4.10, $J(X) = J_0(X)$ and by Definition 3.4 it is easy to see that $\tilde{q} \in I_L[z] = A[z]$ where L is the unique imaginary quadratic field containing A. If q is any element of $A[z] = G$ then by the usual division algorithm (since $\tilde{q}$ is monic) we can write $q = q_1\tilde{q} + q_2$ where $\deg q_2 < n$. Also $q|_J = q_2|_J$, since $\tilde{q}|_J \equiv 0$. Thus in calculating $\operatorname{dist}_J(f, A[z])$ we can restrict our attention to elements of $A[z]$ of degree at most $n - 1$, i.e.,

$$\operatorname{dist}_J(f, A[z]) = \operatorname{dist}_J(f, (A[z])_{n-1}) = \inf_{\deg q \leqslant n-1} \|f - q\|_J.$$

Every element of $(A[z])_{n-1}$ can be expressed in the form $q(z) = a_0 + a_1 z + \cdots + a_{n-1}z^{n-1}$ and then $q(\zeta_i) = a_0 + a_1\zeta_i + \cdots + a_{n-1}\zeta_i^{n-1}$, $1 \leqslant i \leqslant n$.

With these facts in mind we see that the proof of Corollary 10.5 leads to the following.

THEOREM 10.7. *Let X be a Lavrent'ev set with $d(X) < 1$ and $A = I_L$ for some imaginary quadratic field. If $f \in C(X)$ then* $\operatorname{dist}_X(f, A[z])$ *is the minimum of the convex function $\|f - q\|_J$ of q where q runs over all polynomials of the form $q(z) = a_0 + a_1 z + \cdots + a_{n-1}z^{n-1}$, $\{a_0, a_1, \ldots, a_{n-1}\} \subset A$ and subject to the inequalities*

$$|a_0 + a_1\zeta_i + \cdots + a_{n-1}\zeta_i^{n-1}| \leqslant M, \qquad 1 \leqslant i \leqslant n,$$

where M can be taken to be $2\|f\|_J$.

The corresponding result for the real case is the following.

THEOREM 10.8. *Let X be a compact subset of* **R** *with $d(X) < 1$ and $J = J(X, \mathbf{Z}) = \{\chi_1, \ldots, \chi_n\}$. If $f \in C^R(X)$ then* $\operatorname{dist}_X(f, Z[x])$ *is the minimum of the convex function $\|f - q\|_J$ of q where q runs over all polynomials of the form $q(z) = a_0 + a_1 z + \cdots + a_{n-1}z^{n-1}$, $\{a_0, \ldots, a_{n-1}\} \subset \mathbf{Z}$, and subject to the inequalities*

$$|a_0 + a_1\chi_i + \cdots + a_{n-1}\chi_i^{n-1}| \leqslant M, \qquad 1 \leqslant i \leqslant n.$$

Thus in both the real and complex cases the problem reduces to that of minimizing a convex function over a finite subset of a bounded convex subset of an n-dimensional space–a problem which can be solved in a finite number of steps.

It is not difficult to estimate the number of points in this finite subset. Suppose we are in the situation of Theorem 10.8, for simplicity. Then the convex subset in question is the inverse image of the set $|x_i| \leqslant M$ $(1 \leqslant i \leqslant n)$ by a transformation whose determinant is

$$\begin{vmatrix} 1 & \chi_1 & \cdots & \chi_1^{n-1} \\ 1 & \chi_2 & \cdots & \chi_2^{n-1} \\ & & \cdots & \\ 1 & \chi_n & \cdots & \chi_n^{n-1} \end{vmatrix}.$$

The determinant of this (Vandermonde) matrix is well known to be $\Delta = \Pi_{1 \leqslant i < j \leqslant n}(\chi_i - \chi_j)$. Thus the volume of the convex set in question is $2^n M^n (\Pi_{1 \leqslant i < j \leqslant n}(\chi_i - \chi_j))^{-1}$ and the number of lattice points that it contains is approximately the same, for large M. In the real case these lattice points represent the integral polynomials q.

Pointwise approximation. Assume that we are in the situation in Corollary 10.5. Let $f \in C(X)$ but such that f is not interpolable on J by elements of $A[z]$. We know from Proposition 3.7 that f is not A-approximable on J. (Take $X = J$ in the proposition.) Since J is finite it follows that f cannot be approximated in the sense of pointwise convergence on J by elements of $A[z]$. It is possible, however, to approximate pointwise on the complement of J, as follows.

From Corollary 10.5 there exists $q^\circ \in A[z]$ such that

$$\|f - q^\circ\|_J = \operatorname{dist}_J(f, A[z]).$$

Notice that we can assume that $\deg q^\circ < n = \operatorname{card} J$ since dividing by the monic polynomial with the elements of J as simple roots gives a remainder with degree $< n$ and which agrees with q° on J.

THEOREM 10.9. *Under the above hypotheses there exists a sequence $\{q_n\}$ in $A[z]$ such that as $n \to \infty$*

$$q_n(x) \to \begin{cases} f(x), & x \in X \setminus J, \\ q^\circ(x), & x \in J. \end{cases} \tag{5}$$

PROOF. Let $J = \{x_1, \ldots, x_n\}$. It is easy to see that for each j $(1 \leqslant j \leqslant n)$ there is a sequence $\{U_{jk}\}_{k=1}^{\infty}$ of open neighborhoods of x_j satisfying $U_{j1} \supset U_{j2} \supset \ldots$ and $\bigcap_{k=1}^{\infty} U_{jk} = \{x_j\}$. Define $U_k = \bigcup_{j=1}^{n} U_{jk}$. Then $U_1 \supset U_2 \supset \ldots$ and $\bigcap_{k=1}^{\infty} U_k = J$. By Urysohn's lemma there exists a sequence $\{\varphi_n\}$ in $C(X)$ such that

$$0 \leqslant \varphi_n \leqslant 1, \qquad \varphi_n(J) \subset \{0\}, \qquad \varphi_n(X \setminus U_k) \subset \{1\}.$$

Note that $\varphi_n(x) \to 1$ as $n \to \infty$ for all x in $X \setminus J$. Define

$$\psi_n = f\varphi_n + (1 - \varphi_n)q^\circ, \qquad 1 \leqslant n.$$

Then, as $n \to \infty$,

$$\psi_n(x) \to \begin{cases} f(x), & x \in X \setminus J, \\ q^\circ(x), & x \in J. \end{cases} \tag{6}$$

The functions ψ_n are obviously continuous and $\psi_n \equiv q^\circ$ on J; hence they are interpolable on J. Thus by Theorems 4.10 and 4.7, there exist q_n in $A[z]$ with $\|\psi_n - q_n\|_X < 1/n$, $n \geqslant 1$. From this and (6) it is easy to see that (5) holds. □

Using Corollary 10.6 in place of 10.5 and the fundamental results in the real case, the same method of proof gives the following.

COROLLARY 10.10. *Under the hypotheses of Corollary 10.6 there exists a sequence $\{q_n\}$ in $\mathbf{Z}[x]$ such that as $n \to \infty$*

$$q_n(x) \to \begin{cases} f(x), & x \in X \setminus J, \\ q^\circ(x), & x \in J, \end{cases}$$

where q° is an element of $\mathbf{Z}[x]$ satisfying $\|f - q^\circ\|_J = \operatorname{dist}_J(f, \mathbf{Z}[x])$.

It is not difficult to see how to get the analogous result in the general case when we assume that J is finite and X is 1st countable (i.e., every point has a countable neighborhood basis).

PART III: QUANTITATIVE RESULTS

CHAPTER 11

ANALYTIC FUNCTIONS

As usual we let R be any discrete subring of the complex numbers. Throughout this chapter R will always contain the identity.

DEFINITION 11.1. Let f be a complex valued function on a subset X of $\mathbf{C}$. For each n in $\mathbf{N}$ set

$$E_n(f) = \inf_{\deg p \leqslant n} \|f - p\|_X$$

where the p's have arbitrary complex coefficients and

$$E_n^e(f) = \inf_{\deg q \leqslant n} \|f - q\|_X$$

where the coefficients of the q's lie in R.

Clearly

$$\begin{gathered} E_0(f) \geqslant E_1(f) \geqslant \ldots, \\ E_0^e(f) \geqslant E_1^e(f) \geqslant \ldots, \end{gathered} \tag{1}$$

and

$$E_n(f) \leqslant E_n^e(f), \qquad n \in \mathbf{N}. \tag{2}$$

In Part II we were concerned with characterizing those f for which $E_n^e(f) \to 0$ as $n \to \infty$. In Part III we will obtain information about the asymptotic behavior of the sequence $\{E_n^e(f)\}$ as $n \to \infty$. From (2) we have a sort of lower bound on the rate of convergence of $E_n^e(f)$ to zero which we will not mention again.

In the present chapter we will be concerned with the effect on the sequence $\{E_n^e(f)\}$ of the hypothesis that f be analytic. If f is analytic on an interval, then by a well-known theorem of S. N. Bernšteĭn (Lorentz **[66**, 5.5]) there exists $\rho < 1$ such that $E_n(f) = O(\rho^n)$, i.e., there exists a constant C such that $E_n(f) \leqslant C\rho^n$, all n in $\mathbf{N}$. A similar result holds for $E_n^e(f)$. We start with the following.

THEOREM 11.2. *Let q be a monic polynomial in $R[z]$, $0 < \rho_1 < \rho_2$, and $z_1, \ldots, z_k$ the zeros of q. If f is analytic where $|q(z)| < \rho_2$ and $X = \{z: |q(z)| < \rho_1\}$, then $E_n^e(f) = O(\rho^n)$ on X for some $\rho < 1$ if and only if for every positive integer r there is a polynomial q_r in $R[z]$ satisfying*

$$q_r^{(s)}(z_j) = f^{(s)}(z_j), \qquad 1 \leqslant j \leqslant k, 0 \leqslant s \leqslant r. \tag{3}$$

PROOF. First suppose the arithmetic conditions in (3) are satisfied. For the approximating polynomials we may take the interpolating polynomials which are defined as follows. For any positive integer m we let $\tilde{q}_m(z)$ be the (unique) polynomial which interpolates f at the points $z_1, \ldots, z_k$ with multiplicities corresponding to their multiplicities as roots of q^m and which has degree $<$ $m \deg q$. First note that the coefficients of each $\tilde{q}_m$ are in R. Indeed, let μ be the greatest of the multiplicities of the zeros of q. Using the usual division algorithm for polynomials we divide $q_{\mu m}$ by q^m. Since $q_{\mu m}$ and q^m have their coefficients in R by hypothesis, and q^m is monic, it follows that the remainder has its coefficients in R. Also, by the uniqueness of Hermite interpolation polynomials the remainder is identical with $\tilde{q}_m$, as claimed.

Next let γ be given by $|q(z)| = (\rho_1 + \rho_2)/2$. Then the well-known Hermite remainder formula (Walsh [60, Chapter 3]) gives

$$f(z) - \tilde{q}_m(z) = \frac{1}{2\pi i}\int_\gamma \frac{q^m(z)f(\xi)}{q^m(\xi)(\xi - z)}\,d\xi, \qquad |q(z)| \leqslant \rho_1.$$

It follows immediately that

$$\|f - \tilde{q}_m\|_X \leqslant C\rho_0^m \tag{4}$$

where $\rho_0 = 2\rho_1/(\rho_1 + \rho_2) < 1$ and (where L is the length of γ)

$$C = \frac{L}{2\pi}\|f\|_\gamma\Big(\max_{\substack{\xi\in\gamma\\ x\in X}}|\xi - x|\Big)^{-1}.$$

Let $\kappa = \deg q$. Then $\deg \tilde{q}_m < m\kappa$; hence from (4) we have

$$E^e_{m\kappa}(f) \leqslant C\rho_0^m. \tag{5}$$

For any positive integer n we may write $n = m\kappa + t$ where m and t are integers and $0 \leqslant t < \kappa$. Then using (1) and (5) we have

$$E^e_n(f) \leqslant E^e_{m\kappa}(f) \leqslant C\rho_0^m = C\rho_0^{-t/\kappa}(\rho_0^{1/\kappa})^n \leqslant (C\rho_0^{-1})(\rho_0^{1/\kappa})^n;$$

hence the conclusion of the theorem holds with $\rho = \rho_0^{1/\kappa}$.

Conversely, suppose that $E^e_n(f) = O(\rho^n)$ for some $\rho < 1$ and fix a positive integer r. Then for some sequence $\{q_n\}$ of integral polynomials we have $\|f - q_n\|_X = O(\rho^n)$. Applying Cauchy's integral formula in the usual way it is easy to derive from this that, for some constant C,

$$\|f^{(s)} - q_n^{(s)}\|_X \leqslant C\rho^n, \qquad 0 \leqslant s \leqslant r. \tag{6}$$

It follows that, for some n_0,

$$\|f^{(s)} - q_m^{(s)}\|_X < \tfrac{1}{2}, \qquad 0 \leqslant s \leqslant r,$$

whenever $m \geqslant n_0$; hence $\|q_n^{(s)} - q_m^{(s)}\|_X < 1$, $0 \leqslant s \leqslant r$, whenever $m, n \geqslant n_0$. Applying Lemma 3.6 to $q_n^{(s)} - q_m^{(s)}$ we see that

$$q_n^{(s)}(z_j) = q_m^{(s)}(z_j), \qquad 1 \leqslant j \leqslant k, 0 \leqslant s \leqslant r.$$

Since this holds for all $m, n \geqslant n_0$ we have $f^{(s)}(z_j) = q_{n_0}^{(s)}(z_j)$, $1 \leqslant j \leqslant k$, $0 \leqslant s \leqslant r$, as was to be proved. □

It is a curious fact that the arithmetic conditions (3) (i.e., that the coefficients of certain interpolating polynomials lie in R) on the function f impose a limit on the domain of analyticity of f, as we see in the following.

COROLLARY 11.3. *With f and q as in Theorem 11.2 and $f \notin R[z]$ the domain of analyticity of f cannot contain the lemniscate $|q(z)| \leqslant 1$.*

PROOF. We suppose that the domain of analyticity of f does contain $|q(z)| \leqslant 1$. Since the domain of analyticity is open, it must contain $|q(z)| \leqslant 1 + \varepsilon$ for some $\varepsilon > 0$. By the proposition, f is R-approximable on $X = \{z: |q(z)| \leqslant 1\}$. But, as we will see, $d(X) = 1$ which contradicts Theorem 2.11(ii). We can see that $d(X) = 1$ as follows. By definition it suffices to show that a subsequence of the associated sequence of Čebyšev polynomials consists of polynomials with norm 1. But if $\kappa = \deg q$ then $t_{m\kappa}(z, X) = q^m$, $1 \leqslant m$, as follows. It is clear that $|q^m| \equiv 1$ on the boundary of X. If p is a monic polynomial of degree $m\kappa$ with $\|p\|_X < \|q^m\|_X = 1$, then $|p| < |q^m|$ on the boundary of X. Then by Rouché's theorem q^m and $q^m - p$ have the same number of zeros (counting multiplicities) inside X, namely $m\kappa$. But q^m and p are both monic and of degree $m\kappa$, hence $\deg(q^m - p) < m\kappa$. Since the difference has $m\kappa$ zeros, it is the zero polynomial, contradiction. Thus $t_{m\kappa}(z, X) = q^m$ by the definition of Čebyšev polynomials and we are done. □

In the above results we have put rather strong conditions on the domain of analyticity of f. It is in fact enough to assume that f is analytic on an interval $[a, b]$ of the real line, that is, that at every point of $[a, b]$ the function f has a power series expansion with nonzero radius of convergence. We assume for simplicity that we are in the real case. Thus the function to be approximated will be real valued and the coefficients of the approximating polynomials will be from the rational integers.

THEOREM 11.4. *Let f be real valued and analytic on an interval $[a, b]$, $b - a < 4$ and $J([a, b], \mathbf{Z}) = \{z_1, \ldots, z_k\}$. There exist $\rho < 1$ such that $E_n^e(f) = O(\rho^n)$ if and only if for every positive integer r there exists $q_r \in \mathbf{Z}[x]$ such that*

$$q_r^{(s)}(x_j) = f^{(s)}(x_j), \qquad 1 \leqslant j \leqslant k, 0 \leqslant s \leqslant r. \tag{7}$$

PROOF. Since f is analytic we know from the usual proof of Bernšteĭn's theorem (Lorentz [**66**, Chapter 5, Theorem 7]) that f is analytic on some ellipse E with foci at a and b and that $E_n(f) = O(\rho_1^n)$ on this ellipse for some $\rho_1 < 1$. Thus there are polynomials p_n in $\mathbf{R}[x]$ and a constant C_1 such that

$$\|f - p_n\|_E \leqslant C_1\rho_1^n. \tag{8}$$

We first assume that condition (7) holds and show that this leads to $E_n^e(f) = O(\rho^n)$ for some ρ with $\rho_1 < \rho < 1$. The first step is to use (8) to derive a similar estimate for the derivatives of $f - p_n$. Throughout the proof $\|\cdot\| = \|\cdot\|_{[a,b]}$.

Let s be a positive integer and γ the boundary of the ellipse E. By the Cauchy integral formula

$$(f-p_n)^{(s)}(x)=\frac{s!}{2\pi i}\int_\gamma\frac{(f-p_n)(x)}{(\xi-z)^{s+1}}\,d\xi,\qquad a\leqslant x\leqslant b.\tag{9}$$

If we set $\delta=\min_{\xi\in\gamma,z\in[a,b]}|\xi-z|>0$ then we have

$$|(f-p_n)^{(s)}(x)|\leqslant\frac{s!L}{2\pi\delta^{s+1}}\|f-p_n\|_\gamma,\qquad a\leqslant x\leqslant b,\tag{10}$$

where L is the length of γ. Clearly we can assume $\delta<1$ in (10) without loss of generality. Invoking (8) we find that

$$\|(f-p_n)^{(s)}\|_{[a,b]}\leqslant s!C_2\frac{\rho_1^n}{\delta^s}\tag{11}$$

where $C_2=C_1L/2\pi\delta$. For all sufficiently small positive numbers θ we have

$$\rho_1/\delta^\theta<1.\tag{12}$$

We will eventually fix such a θ but assuming we have done so we have

$$\|(f-p_n)^{(s)}\|_{[a,b]}\leqslant s!C_2\left(\frac{\rho_1}{\delta^\theta}\right)^n=s!C_2\rho_2^n,\qquad 1\leqslant s\leqslant[\theta n],$$

where the equality serves to define ρ_2. From (7) we can write

$$|q_r^{(s)}(x_j)-p_n^{(s)}(x_j)|\leqslant C_2s!\rho_2^n,\qquad 1\leqslant j\leqslant k,\tag{13}$$

where $1\leqslant s\leqslant[\theta n]$. The case $s=0$ of (13) follows from (8); hence we have (13) whenever $0\leqslant s\leqslant[\theta n]$.

We know from Proposition 5.9 that there exists a polynomial $\tilde q\in\mathbf{Z}[x]$ with $\|\tilde q\|_{[a,b]}<1$ and $J=J([a,b])=Z_{\tilde q}\cap[a,b]$. Factor $\tilde q$ as follows: $\tilde q=\tilde q_1\tilde q_2$, where $J=Z_{\tilde q_1}$ and $\tilde q_2\neq 0$ on $[a,b]$. Set

$$\tilde{\tilde q}(x)=\prod_{j=1}^k(x-x_j).$$

We can assume without loss of generality that $J\neq\varnothing$ since otherwise $[a,b]$ contains no integers and the conclusion of the theorem follows from Theorem 11.5 below, taking $\rho=\max(\rho_1,\tilde\rho)$ where $\tilde\rho$ is the ρ appearing in Theorem 11.5.

Let $\mu=\deg\tilde q_2$ and pick α $(0<\alpha<1)$ such that on $[a,b]$ we have $\alpha\leqslant|\tilde q_2|\leqslant\alpha^{-1}$. We assume that θ has been chosen small enough that, in addition to (12), we have

$$(1-\alpha^4)^{1/2\mu}\left\{\frac{\alpha(1-\alpha^4)^{(k+\mu)/2\mu}}{\|\tilde{\tilde q}\|}\right\}^{-\theta}<1.\tag{14}$$

Using the usual division algorithm for polynomials we can write, where $r=[\theta n]$,

$$q_r-p_n=\tilde p_n\tilde{\tilde q}^r+\tilde{\tilde p}_n\tag{15}$$

with $\deg\tilde{\tilde p}_n<rk$. Then

$$(q_r-p_n)^{(s)}(x_j)=\tilde{\tilde p}_n(x_j),\qquad 0\leqslant s\leqslant r-1,\quad 1\leqslant j\leqslant k;$$

hence $\tilde{\tilde{p}}_n$ is a Hermite interpolation polynomial to $q_r - p_n$ and by the uniqueness of such polynomials and Gončarov [54, (69), p. 66] it has the form

$$\tilde{\tilde{p}}_n(x) = \sum_{i=1}^{k} \left[\frac{\tilde{\tilde{q}}^r(x)}{(x - x_i)^r} \sum_{j=0}^{r-1} (q_r - p_n)^{(j)}(x_j) \frac{(x - x_i)^j}{j!} \left\{ \frac{(x - x_i)^r}{\tilde{\tilde{q}}^r(x)} \right\}_{(x_i)}^{(r-j-1)} \right]$$

where the notation $\{\varphi\}_{(x_i)}^{(r-j-1)}$ stands for the sum of the terms of the Taylor expansion of φ about x_i whose degree does not exceed $r - j - 1$. The next task is to use (13) to derive an estimate of $\|\tilde{\tilde{p}}_n\|$.

For each i $(1 \leqslant i \leqslant k)$ let γ_i be a circle of radius R_i and centered at x_i such that it does not contain any other point of $J([a, b])$. Without loss of generality we further assume $R_i < 1$ and $\|x - x_i\|_{[a,b]}/R_i \neq 1$ $(1 \leqslant i \leqslant k)$. Set

$$\alpha_{r,i,j} = \left\| \left\{ \frac{(x - x_i)^r}{\tilde{\tilde{q}}^r(x)} \right\}_{(x_i)}^{(r-j-1)} \right\| = \left\| \sum_{s=0}^{r-j-1} a_{s,r,i}(x - x_i)^s \right\|$$

where

$$a_{s,r,i} = \frac{1}{2\pi i} \int_{\gamma_i} \left(\frac{z - x_i}{\tilde{\tilde{q}}(z)} \right)^r \frac{dt}{(t - x_i)^{s+1}}, \qquad 0 \leqslant s \leqslant r - j - 1.$$

Then

$$|a_{s,r,i}| \leqslant \left\| \frac{z - x_i}{\tilde{\tilde{q}}(z)} \right\|_{\gamma_i}^r R_i^{-s}, \qquad 0 \leqslant s \leqslant r - j - 1,$$

and

$$\alpha_{r,i,j} \leqslant \sum_{s=0}^{r-j-1} \left\| \frac{z - x_i}{\tilde{\tilde{q}}(z)} \right\|_{\gamma_i}^r (\|x - x_i\|/R_i)^s.$$

Setting $\beta_i = \|x - x_i\|/R_i$ we have

$$\alpha_{r,i,j} \leqslant \left\| \frac{z - x_i}{\tilde{\tilde{q}}(z)} \right\|_{\gamma_i}^r \frac{\beta_i^{r-j} - 1}{\beta_i - 1} = O\left(\left\| \frac{z - x_i}{\tilde{\tilde{q}}(z)} \right\|_{\gamma_i}^r \beta_i^{r-j} \right).$$

Consequently (recall $R_i < 1$)

$$\sum_{j=0}^{r-1} \|x - x_i\|^j \alpha_{r,i,j} = O\left(r \left\| \frac{z - x_i}{\tilde{\tilde{q}}(z)} \right\|_{\gamma_i}^r \beta_i^r \right).$$

From (13) and (15)

$$\|\tilde{\tilde{p}}_n\| = \rho_2^n \sum_{i=1}^{k} \left\| \frac{\tilde{\tilde{q}}(z)}{z - x_i} \right\|^r O\left(r \left\| \frac{z - x_i}{\tilde{\tilde{q}}(z)} \right\|_{\gamma_i}^r \beta_i^r \right)$$

$$= O\left[\rho_2^n k r \left\{ \max_{1 \leqslant i \leqslant k} \left\| \frac{\tilde{\tilde{q}}(z)}{z - x_i} \right\|_{[a,b]} \left\| \frac{z - x_i}{\tilde{\tilde{q}}(z)} \right\|_{\gamma_i} \beta_i \right\}^r \right].$$

Let K denote the expression within the braces. Then we have $\|\tilde{\tilde{p}}_n\| = O(\rho_2^n r K^r) = O(\rho_2^n (K + 1)^r)$. Since K does not depend on n we can assume that θ has been chosen so small that, in addition to (12) and (14), we have $\rho_3 = \rho_2(K+1)^\theta < 1$. Then (recall $r = [\theta n]$)

$$\|\tilde{\tilde{p}}_n\| = O(\rho_3^n). \tag{16}$$

Combining (8), (15) and (16) we have

$$\|f - q_r - \tilde{\tilde{q}}^r \tilde{p}_n\| = O(\rho_1^n + \rho_3^n). \tag{17}$$

It remains to approximate the term $\tilde{\tilde{q}}^r \tilde{p}_n$ by an integral polynomial. We first approximate $\tilde{p}_n$ by an integral polynomial plus a polynomial which is divisible by $\tilde{q}_2^r$. This is accomplished by expanding $\tilde{p}_n$ in terms of powers of $\tilde{q}_2$ and then approximating the "nonintegral parts" of lower powers of $\tilde{q}_2$ by polynomials in higher powers as follows.

First note that for any positive integers m, j

$$\left|x^j - \left(x^j - x^j(1-\alpha^2x^2)^m\right)\right| \leqslant \alpha^{-j}(1-\alpha^4)^m, \qquad \alpha \leqslant |x| \leqslant \alpha^{-1};$$

hence we have

$$\begin{aligned}\left\|\tilde{q}_2^j - \left(\tilde{q}_2^j - \tilde{q}_2^j(1-\alpha^2\tilde{q}_2^2)^m\right)\right\| &= \left\|\tilde{q}_2^j - \left(m\alpha^2\tilde{q}_2^{j+2} + \cdots + d\tilde{q}_2^{j+2m}\right)\right\| \\ &\leqslant \alpha^{-j}(1-\alpha^4)^m. \end{aligned} \tag{18}$$

By Lemma 4.4 we can write

$$\tilde{p}_n = \sum_{j=0}^{N_i} A_j \tilde{q}_2^j \tag{19}$$

where $\deg A_j < \mu$ $(= \deg \tilde{q}_2)$, $0 \leqslant j \leqslant N_1$, and $N_1 = [\deg \tilde{p}_n/\mu] \leqslant [(n-kr)/\mu]$. If we set

$$m = \left[\frac{[(n-kr)/\mu] - r + 1}{2}\right]$$

then by a straightforward calculation

$$\frac{n}{2\mu} - \frac{r}{2}\left(\frac{k}{\mu} + 1\right) - \frac{1}{2} \leqslant m \leqslant \frac{[(n-kr)/\mu] - r + 1}{2}. \tag{20}$$

For each j $(0 \leqslant j \leqslant r-1)$ we denote by $[A_j]$ the polynomial obtained from A_j by replacing each coefficient by its integral part and set $(A_j) = A_j - [A_j]$. Then for $0 \leqslant j \leqslant r-1$ we have

$$\|(A_j)\| \leqslant C = \sum_{i=0}^{\mu-1} \|x\|^i. \tag{21}$$

We are now prepared to carry out the approximation of $\tilde{p}_n$. Starting with (19)

$$\tilde{p}_n = [A_0] + \sum_{j=1}^{N_1} A_{1,j}\tilde{q}_2^j + \varepsilon_0$$

$$= [A_0] + [A_1]\tilde{q}_2 + \sum_{j=2}^{N_1} A_{2,j}\tilde{q}_2^j + \varepsilon_0 + \varepsilon_1$$

$$\cdots$$

$$= \sum_{j=0}^{r-1} [A_j]\tilde{q}_2^j + \sum_{j=r}^{N_1} A_{r,j}\tilde{q}_2^j + \sum_{j=0}^{r-1} \varepsilon_j \tag{22}$$

where in each step the factor $\tilde{q}_2^j$ in $(A_{i,j})\tilde{q}_2^j$ is replaced by its approximation in (18). The upper limit of summation of the second term remains the same by virtue of the right-hand side of (20). From (18) and (21) we have

$$\sum_{j=0}^{r-1} \varepsilon_j \leqslant C(1-\alpha^4)^m \sum_{j=0}^{r-1} \alpha^{-j}$$

$$= C(1-\alpha^4)^m \frac{\alpha^{-r}-1}{\alpha-1} = O\big((1-\alpha^4)^m \alpha^{-r}\big).$$

But by (20),

$$O\big((1-\alpha^4)^m\alpha^{-r}\big) = O\big((1-\alpha^4)^{(n/2\mu - r(k/\mu+1)/2-1/2)}\alpha^{-r}\big)$$

$$= O\big((1-\alpha^4)^{(n/2\mu-\theta n(k/\mu+1)/2)}\alpha^{-\theta n}\big).$$

Thus from (22),

$$\tilde{\tilde{q}}^r\tilde{p}_n = \tilde{\tilde{q}}^r \sum_{j=0}^{r-1} [A_j]\tilde{q}_2^j + \tilde{\tilde{q}}^r \sum_{j=r}^{N_1} A_{r,j}\tilde{q}_2^j + O(\rho^4) \tag{23}$$

where $\rho_4 = (1-\alpha^4)^{1/2\mu}\{\alpha(1-\alpha^4)^{(k+\mu)/2\mu}/\|\tilde{\tilde{q}}\|\}^{-\theta} < 1$ by (14).

It remains to approximate the middle term π_n, say, of the right-hand side of (23). Since $\tilde{\tilde{q}}$ has the same zeros as $\tilde{q}_1$ but with multiplicity 1 there is a positive integer l such that $\tilde{q}_1$ divides $\tilde{\tilde{q}}^l$. It follows that $\tilde{q}^{[r/l]}$ divides $(\tilde{\tilde{q}}\tilde{q}_2)^r$. Setting $\theta_1 = \theta/l$ we have $0 < \theta_1 < 1$ and $[\theta_1 n] = [\theta n/l] \leqslant [\theta n]/l = r/l$; hence $[\theta_1 n] \leqslant [r/l]$. Thus $\tilde{q}^{[\theta_1 n]}$ divides π_n and by Lemma 4.4 we can write

$$\pi_n = \sum_{j=[\theta_1 n]}^{N_2} B_j\tilde{q}^j$$

where $\deg B_j < \deg \tilde{q}$, all j; hence $\|(B_j)\| \leqslant C_3$ for some constant C_3 just as for $\|(A_j)\|$. Thus

$$\pi_n = \sum_{j=[\theta_1 n]}^{N_2} [B_j]\tilde{q}^j + \sum_{j=[\theta_1 n]}^{N_2} (B_2)\tilde{q}^j$$

and

$$\left\| \sum_{j=[\theta_1 n]}^{N_2} (B_j)\tilde{q}^j \right\| \leqslant C_3 \sum_{j=[\theta_1 n]}^{\infty} \|\tilde{q}\|^j = C_3 \frac{\|\tilde{q}\|^{[\theta_1 n]}}{1 - \|\tilde{q}\|}$$

$$\leqslant \frac{C_3}{(1 - \|\tilde{q}\|)\|\tilde{q}\|} \|\tilde{q}\|^{\theta_1 n} = O(\rho_5^n) \tag{24}$$

where $\rho_5 = \|\tilde{q}\|^{\theta_1} < 1$.

From (17), (23) and (24) we have

$$\left\| f - \left(q_r + \tilde{q}^r \sum_{j=0}^{r-1} [A_j] \tilde{q}_2^j + \sum_{j=[\theta_1 n]}^{N_2} [B_j] \tilde{q}^j \right) \right\| \leqslant O(\rho^n)$$

where $\rho = \max\{\rho_1, \rho_3, \rho_4, \rho_5\} < 1$. This completes the proof of one direction of the theorem, since $q_2 \in \mathbf{Z}[x]$.

Conversely, suppose that $E_n^e(f) = O(\rho^n)$ for some $\rho < 1$. Then for some sequence of integral polynomials $\{\tilde{q}_n\}$ we have $\|f - \tilde{q}_n\|_{[a,b]} = O(\rho^n)$. As in the usual proof of Bernšteĭn's theorem (Lorentz [**66**, Chapter 5, Theorem 7]) we can see that the convergence actually takes place on some ellipse with a and b as foci. Proceeding as in the proof of Theorem 11.2, we see that (7) holds for some element of the sequence $\{\tilde{q}_n\}$ in place of q_r. □

The following applies to more general functions than the analytic ones. It is included in this chapter since it is needed to complete the proof of Theorem 11.4. Although it is formulated for an interval of the form $[\delta, 1 - \delta]$, $\delta < \frac{1}{2}$, it holds on any (closed) interval not containing an integer since we can translate by an integral amount and then integral polynomials go into integral polynomials.

THEOREM 11.5. *Let f be a real valued function defined and continuous on the interval* $[\delta, 1 - \delta]$, $0 < \delta < \frac{1}{2}$. *Then for any positive integer n*

$$E_n^e(f) \leqslant E_n(f) + 2n\rho^n$$

where $\rho = \max\{\frac{1}{2}, (1 - 2\delta)/(2\sqrt{\delta(1 - \delta)} + 1)\} < 1$.

Note that by an easy calculation

$$\rho = \begin{cases} \dfrac{1}{2}, & \dfrac{1}{10} \leqslant \delta < \dfrac{1}{2}, \\[2ex] \dfrac{1 - 2\delta}{2\sqrt{\delta(1 - \delta)} + 1}, & 0 < \delta \leqslant \dfrac{1}{10}. \end{cases}$$

PROOF. Let p_n be the polynomial of degree at most n which satisfies ($\|\cdot\| = \|\cdot\|_{[\delta, 1-\delta]}$):

$$\|f - p_n\| = E_n(f).$$

Using Lemma 4.4 we can write

$$p_n(x) = \sum_{s=0}^{[n/2]} A_s x^s (1 - x)^s \tag{25}$$

where the A_s ($0 \leqslant s \leqslant [n/2]$) are linear polynomials. It is thus easy to see that for each such A_s there is an integral linear polynomial $[A_s]$ satisfying $\|A_s - [A_s]\| < \frac{1}{2}$. The fundamental idea of the proof is to approximate, successively, the nonintegral parts of the terms in (25) by terms of higher degree, as follows. Suppose $n \geqslant 4$ and let $m = n - 1 - (1 + (-1)^n)/2$. Then m is even. If we set

$$C_m(t) = \tfrac{1}{2}\left\{\left(t + \sqrt{t^2 - 1}\,\right)^m + \left(t - \sqrt{t^2 - 1}\,\right)^m\right\} \tag{26}$$

and

$$T_m(x) = C_m((1 - 2x)/(1 - 2\delta))$$

then C_m coincides with the Čebyšev polynomial on $[-1, 1]$ of degree m, hence contains terms of even degree only and $|T_m(x)| = |C_m(t)| \leqslant 1$ (Lorentz [**66**, Chapter 2, §7]). It follows that $T_m(x) - T_m(0)$ has both $x = 0$ and $x = 1$ as roots, hence is divisible by $x(1 - x)$. Thus for some polynomial $\tilde{T}_{m-2}$ of degree $m - 2$ we have $T_m(x) = T_m(0) - x(1 - x)T_m(0)\tilde{T}_{m-2}(x)$; hence

$$1 \geqslant |T_m(0) - x(1 - x)T_m(0)\tilde{T}_{m-2}(x)|, \qquad \delta \leqslant x \leqslant 1 - \delta. \tag{27}$$

Notice also that from (26),

$$T_m(0) = \frac{1}{2}\left\{\left[\frac{1}{1 - 2\delta} + \sqrt{\frac{1}{(1 - 2\delta)^2} - 1}\,\right]^m + \left[\frac{1}{1 - 2\delta} - \sqrt{\frac{1}{(1 - 2\delta)^2} - 1}\,\right]^m\right\}$$

$$\geqslant \frac{1}{2}\left[\frac{1}{1 - 2\delta} + \sqrt{\frac{1}{(1 - 2\delta)^2} - 1}\,\right]^m \geqslant \frac{1}{2}\rho^{-m} > 0. \tag{28}$$

Using this and (27), multiplied through by $x^k(1 - x)^k$, we see that

$$\|x^k(1 - x)^k - x^{k+1}(1 - x)^{k+1}\tilde{T}_{m-2k-2}(x)\| \leqslant \frac{\|x^k(1 - x)^k\|}{T_{m-2k}(0)} = \frac{1}{2^{2k}T_{m-2k}(0)} \leqslant 2\rho^m \tag{29}$$

for $0 \leqslant k \leqslant ([n/2] - 2)$. Thus (setting $(A_0) = A_0 - [A_0]$):

$$A_0 = [A_0] + (A_0) = [A_0] + (A_0)x(1 - x)\tilde{T}_{m-2}(x) + \varepsilon_0(x)$$

where

$$\begin{aligned}\|\varepsilon_0(x)\| &= \|(A_0) - (A_0)x(1 - x)\tilde{T}_{m-2}(x)\| \\ &\leqslant \|(A_0)\|\,\|1 - x(1 - x)\tilde{T}_{m-2}(x)\| \leqslant \tfrac{1}{2}2\rho^m = \rho^m.\end{aligned}$$

Consequently, starting with (25) we can write

$$\begin{aligned}p_n(x) &= [A_0] + (A_0)x(1 - x)\tilde{T}_{m-2}(x) + \sum_{s=1}^{[n/2]} A_s x^s(1 - x)^s + \varepsilon_0(x) \\ &= [A_0] + \sum_{s=1}^{[n/2]} A_{1,s}x^s(1 - x)^s + \varepsilon_0(x)\end{aligned}$$

where $A_{1,s}$ is a linear polynomial ($1 \leqslant s \leqslant [n/2]$). We next make the same type of approximation to the term $A_{1,1}x(1-x)$, using (29) with $k=1$, obtaining

$$p_n(x)=[A_0]+[A_{1,1}]x(1-x)+\sum_{s=2}^{[n/2]} A_{2,s}x^s(1-x)^s+\varepsilon_0(x)+\varepsilon_1(x).$$

Continuing in this way we obtain the presentation ($A_{0,0}=A_0$):

$$p_n(x)=\sum_{s=0}^{[n/2]-2}[A_{s,s}]x^s(1-x)^s+\sum_{s=[n/2]-1}^{[n/2]} B_s x^s(1-x)^s+\sum_{s=0}^{[n/2]-2}\varepsilon_s(x)$$

where the B_s's are linear. But $\|\Sigma_{s=0}^{[n/2]-2}\varepsilon_s(x)\|\leqslant([n/2]-1)\rho^m$ and

$$\|(B_s)x^s(1-x^s)\|\leqslant\|(B_s)\|\,\|x^s(1-x)^s\|\leqslant\frac{1}{2\cdot 2^{2s}}\leqslant\frac{1}{2}\rho^{2s},$$

$[n/2]-1\leqslant s\leqslant[n/2]$; hence for

$$q_n(x)=\sum_{s=0}^{[n/2]-2}[A_{s,s}]x^s(1-x)^s+\sum_{s=[n/2]-1}^{[n/2]}[B_s]x^s(1-x)^s$$

we have

$$\|p_n-q_n\|\leqslant([n/2]-1)\rho^m+3\rho^{2[n/2]}.$$

Considering the cases of even n and odd n separately and recalling that $n\geqslant 4$, it is straightforward to calculate from this that $\|p_n-q_n\|\leqslant 2n\rho^n$. In case $n\leqslant 3$ we have $p_n(x)=A_0+A_1x(1-x)$ and setting $q_n=[A_0]+[A_1]x(1-x)$ we have

$$|p_n(x)-q_n(x)|=|(A_0)+(A_1)x(1-x)|\leqslant\frac{1}{2}+\frac{1}{2}\cdot\frac{1}{4}=\frac{5}{8},$$

but $2n\rho^n\geqslant n2^{1-n}\geqslant 3/4$ $(1\leqslant n\leqslant 3)$. □

REMARK 11.6. It is sometimes useful to know that the arithmetic conditions in Theorems 11.2 and 11.4 are equivalent to requiring that for every positive integer r the Hermite interpolation polynomial h_r defined by $\deg h_r<(r+1)k$ and

$$h_r^{(s)}(z_j)=f^{(s)}(z_j),\qquad 1\leqslant j\leqslant k,\ \ 0\leqslant s\leqslant r, \tag{30}$$

has only integral coefficients. We thus have a constructive criterion for determining the approximability of f. We can see this as follows. Suppose that q_r is an integral polynomial which interpolates as in (30). If we set $\tilde{q}(z)=\Pi_{j=1}^k(z-z_j)$ then $\tilde{q}$ is an integral polynomial since it is a product of minimal polynomials of algebraic integers. Then by the division algorithm $q_r=p\tilde{q}^{r+1}+t$ where p and t have integral coefficients and $\deg t<(r+1)k$. But it is clear that t also interpolates q_r, hence f as in (30). By the uniqueness of the Hermite interpolation polynomial, then, $t=h_r$ and we are done.

EXAMPLE 11.7. In the special case of Theorem 11.4 where $[a,b]=[0,1]$ the arithmetic conditions (7) become simply that for every nonnegative integer j both

$$f^{(s)}(0)/s!\ \text{ and }\ f^{(s)}(1)/s!\ \text{ are integers.} \tag{31}$$

Indeed, as we saw in the proof, if condition (7) holds, then a sequence of integral polynomials converges uniformly to f on some ellipse with 0 and 1 as foci and (31) follows by Proposition 7.10. Conversely, by Lemma 4.4 we can represent the polynomials $\{q_r\}$ in (7) in the form

$$q_r = \sum_{j=0}^{r} A_j \tilde{\tilde{q}}^j \tag{32}$$

where $\tilde{\tilde{q}}(x) = x(1 - x)$ and $\deg A_j < 2$ $(1 \leqslant j \leqslant r)$. We have assumed $\deg q_r < 2(r + 1)$ without loss of generality by Remark 11.6. Notice that the A_j's do not depend on q_r as follows. As in the argument in Remark 11.6, A_0 is the interpolation polynomial to q_r on J, but $q_r \equiv f$ on J; A_1 is the interpolating polynomial to $(q_r - A_0)/\tilde{\tilde{q}}$ on J, but $(q^r - A_0)/\tilde{\tilde{q}} = (f - A_0)/\tilde{\tilde{q}}$ on J; etc. Clearly, we can write the A_j's in the form $A_j = a_j x + b_j(1 - x)$. We wish to establish from the assumption (31) that the q_r's are integral; hence it suffices to show that the a_j's and b_j's are integers. We proceed by induction on j. Clearly $a_0 = f(1)$ and $b_0 = f(0)$. Suppose $A_0, \ldots, A_{s-1}$ are integral. Then from (32) (with $r > s$)

$$f^{(s)}(0)/s! = q_r^{(s)}(0)/s! = \frac{1}{s!}\left(\sum_{j=0}^{r} A_j \tilde{\tilde{q}}^j\right)^{(s)}\Bigg|_{x=0}$$

$$= \frac{1}{s!}\left(\sum_{j=0}^{s-1} A_j \tilde{\tilde{q}}^j\right)^{(s)}\Bigg|_{x=0} + \frac{1}{s!}(A_s \tilde{\tilde{q}}^s)^{(s)} = \frac{1}{s!}\left(\sum_{j=0}^{s-1} A_j \tilde{\tilde{q}}^j\right)^{(s)}\Bigg|_{x=0} + b_s.$$

Thus b_s is the difference between $f^{(s)}(0)/s!$ and the sth coefficient of an integral (by the induction hypothesis) polynomial, hence is an integer. Similarly,

$$a_s = \frac{f^{(s)}(1)}{s!} - \frac{1}{s!}\left(\sum_{j=0}^{s} A_j \tilde{\tilde{q}}^j\right)^{(s)}\Bigg|_{x=1}$$

and the last term is the sth coefficient of the Taylor series of an integral polynomial at $x = 1$, hence an integer (substitute $(1 - (1 - x))$ for x).

CHAPTER 12

FINITELY DIFFERENTIABLE FUNCTIONS

In this chapter we will consider the degree of approximation of real valued functions defined on intervals (of length less than 4) and which possess finitely many derivatives. The coefficients of the approximating polynomials will come from the ring of rational integers $\mathbf{Z}$.

For any nonnegative integer r let $C^r([a, b])$ denote the real valued functions on $[a, b]$ which possess a continuous rth derivative ($C^0([a, b]) = C([a, b], \mathbf{R})$ by definition). Of course $C^r([a, b]) \subset C^\nu([a, b])$, $0 \leqslant \nu \leqslant r$. If g is a function on $[a, b]$ then we denote by $\omega(g, \cdot)$ the modulus of continuity of g, i.e., the function defined on $[0, b - a]$ by

$$\omega(g, h) = \sup\{|g(x) - g(y)| : |x - y| \leqslant h \text{ and } x, y \in [a, b]\}.$$

We take for granted the elementary properties of moduli of continuity (Lorentz [**66**, Chapter 3, §5]).

We suppose throughout the chapter that, unless otherwise indicated, $f \in C^r([a, b])$, $b - a < 4$, $\omega(\cdot) = \omega(f^{(r)}, \cdot)$, and $J([a, b]) = \{x_1, \ldots, x_k\}$. The fundamental result is the following. Its conclusion can be strengthened in a special case as we will see in Theorem 12.10.

THEOREM 12.1. *There exist a number C depending on f, r, and $[a, b]$ but not on n or x and a sequence $\{q_n\}$ of polynomials in $\mathbf{Z}[x]$, $\deg q_n \leqslant n$, such that for $a \leqslant x \leqslant b$, $0 \leqslant \nu \leqslant r$, and any positive integer n we have*

$$|f^{(\nu)}(x) - q_n^{(\nu)}(x)| \leqslant C\left(\max\left\{\frac{\sqrt{(x-a)(b-x)}}{n}, \frac{1}{n^2}\right\}\right)^{r-\nu} \times \omega\left(\max\left\{\frac{\sqrt{(x-a)(b-x)}}{n}, \frac{1}{n^2}\right\}\right) \tag{1}$$

if and only if there is an element q of $\mathbf{Z}[x]$ satisfying

$$q^{(\nu)}(x_j) = f^{(\nu)}(x_j), \qquad 0 \leqslant \nu \leqslant r, \quad 1 \leqslant j \leqslant k. \tag{2}$$

Just as in Remark 11.6 we can see that the criterion for approximability can be replaced by the equivalent condition that the Hermite interpolation polynomial satisfying (2) has only integral coefficients.

The necessity of the condition (2) can be seen as follows. Since $\omega(h) \to 0$ as $h \to 0$ we see that (1) implies the uniform convergence of $q_n^{(\nu)}$ to $f^{(\nu)}$ on $[a, b]$, $0 \leqslant \nu \leqslant r$. Thus for some n_0 ($\|\cdot\| = \|\cdot\|_{[a,b]}$), $\|q_n^{(\nu)} - f^{(\nu)}\| < \frac{1}{2}$, $0 \leqslant \nu \leqslant r$, $n \geqslant n_0$; hence

$$\|q_n^{(\nu)} - q_m^{(\nu)}\| < 1, \qquad 0 \leqslant \nu \leqslant r, \quad n, m \geqslant n_0.$$

It follows from Lemma 3.6 that, for $m, n \geqslant n_0$, $q_n^{(\nu)}(x_j) = q_m^{(\nu)}(x_j)$, $0 \leqslant \nu \leqslant r$, $1 \leqslant j \leqslant k$; hence

$$q_{n_0}^{(\nu)}(x_j) = f^{(\nu)}(x_j), \qquad 0 \leqslant \nu \leqslant r, \quad 1 \leqslant j \leqslant k.$$

The remainder of the proof will follow after a series of auxiliary results. A rough outline of the proof is as follows. First we will show that this kind of approximation is possible by polynomials with arbitrary (real) coefficients. These polynomials will then be expanded in terms of the powers of $\tilde{q}(x)$, an integral polynomial with $\|\tilde{q}\|_{[a,b]} < 1$. The (polynomial) coefficients in this expansion are then approximated by integral polynomials.

The first result is an interesting extension by A. O. Gel'fond of a theorem of A. F. Timan. It does not appear in the standard treatises; hence we include a proof. For any positive integer n let

$$\Delta_n(x) = \max\left\{\frac{\sqrt{1 - x^2}}{n}, \frac{1}{n^2}\right\}, \qquad |x| \leqslant 1.$$

THEOREM 12.2. *If $f \in C^r - ([-1, 1])$ and $\omega(\cdot) = \omega(f^{(r)}, \cdot)$ then for any positive integer n there is a polynomial $p_n(x)$, $\deg p_n \leqslant n$, such that for $-1 \leqslant x \leqslant 1$ and $0 \leqslant \nu \leqslant r$*

$$|f^{(\nu)}(x) - p_n^{(\nu)}(x)| \leqslant C_r \Delta_n^{r-\nu}(x)\omega(\Delta_n(x)) \tag{3}$$

where C_r does not depend on n or x.

What we will actually need from Theorem 12.2 is the following slightly stronger result. Let

$$\tilde{\Delta}_n(x) = \max\left\{\frac{\sqrt{(x - a)(b - x)}}{n}, \frac{1}{n^2}\right\}.$$

COROLLARY 12.3. *Let $f \in C^r([a, b])$, $\omega(\cdot) = \omega(f^{(r)}, \cdot)$, and $\{x_1, \ldots, x_k\} \subset [a, b]$. Then there exists a sequence $\{p_n\}$ of polynomials with real coefficients, $\deg p_n \leqslant n$, such that*

$$|f^{(\nu)}(x) - p_n^{(\nu)}(x)| \leqslant C_r \tilde{\Delta}_n^{r-\nu}(x)\omega(\tilde{\Delta}_n(x)), \qquad a \leqslant x \leqslant b, \tag{4}$$

where C_r does not depend on n or x and

$$f^{(\nu)}(x_j) = p_n^{(\nu)}(x_j) \tag{5}$$

whenever $0 \leqslant \nu \leqslant r$, $1 \leqslant j \leqslant k$, and $n \geqslant (r + 1)k$.

We first assume the validity of Theorem 12.2 and prove the corollary. Using the standard (linear) change of variables it is straightforward to derive from Theorem 12.2 the existence of a sequence $\{p_n\}$ satisfying (4). The Hermite interpolation polynomial $\tilde{p}_n$ defined by

$$\tilde{p}_n^{(\nu)}(x_j) = f^{(\nu)}(x_j) - p_n^{(\nu)}(x_j), \qquad 0 \leqslant \nu \leqslant r, \quad 1 \leqslant j \leqslant k,$$

has degree $< (r + 1)k$; hence for $n \geqslant (r + 1)k$ we can add it to p_n without invalidating the condition $\deg p_n \leqslant n$. Also, each $\tilde{p}_n$ is a linear combination of certain polynomials not depending on n with coefficients $f^{(\nu)}(x_j) - p_n^{(\nu)}(x_j)$ as we see from Gončarov [**54**, (69), p. 66] for example. Thus for each ν we have from (4) that

$$|\tilde{p}_n^{(\nu)}(x)| = O(\tilde{\Delta}_n^{r-\nu}(x)\omega(\tilde{\Delta}_n(x))), \qquad a \leqslant x \leqslant b.$$

Thus we can add the $\tilde{p}_n$'s ($n \geqslant (r + 1)n$) to the p_n's and both conditions (4) and (5) will be satisfied (with a possible change in C_r). □

PROOF (OF THEOREM 12.2). We proceed by induction on r. The case $r = 0$ is Lorentz [**66**, Chapter 5, Theorem 1] whose notation we adopt, by and large. Assume the validity of the theorem with r replaced by $r - 1$. Applying it to f' we obtain, for $|x| \leqslant 1$,

$$|f^{(\nu+1)}(x) - \tilde{p}_n^{(\nu)}(x)| \leqslant C_{r-1}\Delta_n^{r-1-\nu}(x)\omega(\Delta_n(x)), \qquad 0 \leqslant \nu \leqslant r - 1, \tag{6}$$

for some sequence of polynomials $\tilde{p}_n$, $\deg p_n \leqslant n$, and where $\omega(\cdot) = \omega((f')^{(r-1)}, \cdot) = \omega(f^{(r)}, \cdot)$. If we let $\tilde{\tilde{p}}_{n+1}$ be an indefinite integral of $\tilde{p}_n$ then we have

$$|f^{(\nu)}(x) - \tilde{\tilde{p}}_{n+1}^{(\nu)}(x)| \leqslant C_{r-1}\Delta_n^{r-\nu}(x)\omega(\Delta_n(x)), \qquad 1 \leqslant \nu \leqslant r.$$

Since $\Delta_n(x) \leqslant 4\Delta_{n+1}(x)$, hence $\omega(\Delta_n(x)) \leqslant 4\omega(\Delta_{n+1}(x))$, we can write

$$|f^{(\nu)}(x) - \tilde{\tilde{p}}_{n+1}^{(\nu)}(x)| \leqslant C_{r-1}\Delta_{n+1}^{r-\nu}(x)\omega(\Delta_{n+1}(x)), \qquad 1 \leqslant \nu \leqslant r, \tag{7}$$

where C_{r-1} has been multiplied by an appropriate power of 4 but we write it the same way to conserve notation.

Next let n be any positive integer and set

$$k_{nr}(u) = \lambda_{n'r}^{-1}\left(\frac{\sin(n'u/2)}{\sin(u/2)}\right)^{2r}, \qquad n' = [n/r] + 1$$

where $\lambda_{n'r}$ is a constant, adjusted so as to have $\int_{-\pi}^{\pi} k_{nr}(u)\,du = 1$. Further define $t = \arccos x$ and for any positive integer n set $g = f - \tilde{\tilde{p}}_{n+1}$,

$$J_n(x) = \int_{-\pi}^{\pi} g(\cos(t + u))k_{n,r+2}(u)\,du, \tag{8}$$

and $p_{n+1} = \tilde{\tilde{p}}_{n+1} + J_n$. We will show that for $x \in [-1, 1]$,

$$|f(x) - p_{n+1}(x)| \leqslant \tilde{C}_r\Delta_{n+1}^r(x)\omega(\Delta_{n+1}(x)) \tag{9}$$

and

$$|J_n^{(\nu)}(x)| \leqslant \tilde{\tilde{C}}_r\Delta_{n+1}^{r-\nu}(x)\omega(\Delta_{n+1}(x)), \qquad 1 \leqslant \nu \leqslant r. \tag{10}$$

By standard arguments concerning convolution with powers of the Fejér kernel (Lorentz [**66**, Chapter 5]) J_n is a polynomial of degree $\leqslant n$. It follows that $\deg p_{n+1} \leqslant n + 1$. In view of (7), (9), and (10), the lemma will be proved for $n \geqslant 2$ and the general case ($n \geqslant 1$) follows by a change in the constant C_r.

We first consider (9). Define a function ϕ by

$$\phi(u) = C_{r-1}u^{r-1}\omega(u), \qquad u \geqslant 0.$$

Then ϕ satisfies Lorentz [**66**, 5.1(7)] with $m = r$; hence, applying Lorentz [**66**, Chapter 5, Lemma 3] to the case $\nu = 0$ of (6), we immediately obtain (9).

To establish (10) we proceed as follows. We have, starting with (8),

$$\begin{aligned} J_n'(x) &= \frac{d}{dx}\int_{-\pi}^{\pi} g(\cos(t+u))k_{n,r+2}(u)\,du \\ &= \frac{d}{dx}\int_{-\pi}^{\pi} g(\cos(u))k_{n,r+2}(u-t)\,du \\ &= \int_{-\pi}^{\pi} g(\cos(u))\frac{1}{\sin t}k'_{n,r+2}(u-t)\,du \\ &= \int_{-\pi}^{\pi} \frac{g(\cos(u+t))}{\sin t}k'_{n,r+2}(u)\,du \\ &= \int_{0}^{\pi} \frac{g(\cos(u+t)) - g(\cos(u-t))}{\sin t}k'_{n,r+2}(u)\,du \\ &= \int_{0}^{\pi} -g'(\cos(\xi))(2\sin u)k'_{n,r+2}(u)\,du \end{aligned} \tag{11}$$

where, in the last step, we have used the mean value theorem for g; hence $\xi = t + \theta u$ for some θ with $0 \leqslant |\theta| < 1$ ($\cos\xi$ is an even function). But (writing $a_n \approx b_n$ to mean $m \leqslant a_n/b_n \leqslant M$ for some constants $M, m > 0$)

$$\begin{aligned} k'_{n,r+2}(u) &= (2r+4)\lambda_{n',r+2}^{-1}\left(\frac{\sin(n'u/2)}{\sin(u/2)}\right)^{2r+3}\left(\frac{\sin(n'u/2)}{\sin(u/2)}\right)' \\ &= O\left\{(2r+4)\lambda_{n',r+1}^{-1}\left(\frac{\sin(n'u/2)}{\sin(u/2)}\right)^{2r+2}\left(\frac{\sin(n'u/2)}{n'\sin(u/2)}\right)'\right\} \end{aligned}$$

where we have used the fact that $\lambda_{n',r} \approx (n')^{2r-1}$ (Lorentz [**66**, Chapter 4, §3(2)]) and $|(\sin(n'u/2))/(n'\sin(u/2))| \leqslant 1$ which is obvious. It is easily checked that

$$|\sin u|\left|\frac{d}{du}\left(\frac{\sin(n'u/2)}{n'\sin(u/2)}\right)\right|$$

is bounded by an absolute constant; hence from (11)

$$|J_n'(x)| \leqslant M_r\int_0^{\pi}|g'(\cos(\xi))|k_{n,r+1}(u)\,du. \tag{12}$$

Now by (6) with $\nu = 0$ we have

$$|g'(\cos(\xi))| \leqslant C_{r-1}\Delta_n^{r-1}(t+\theta u)\omega(\Delta_n(t+\theta u)). \tag{13}$$

Also

$$\Delta_n(\cos(t+\theta u)) = \max\left\{\frac{\sqrt{1-\cos^2(t+\theta u)}}{n}, \frac{1}{n^2}\right\}$$
$$= \max\left\{\frac{|\sin(t+\theta u)|}{n}, \frac{1}{n^2}\right\}$$
$$\leqslant \max\left\{\frac{|\sin t|}{n} + \frac{|u|}{n}, \frac{1}{n^2}\right\} \leqslant \delta_n(t) + \frac{|u|}{n}$$

where by definition $\delta_n(t) = \max\{|\sin t|/n, 1/n^2\}$. Thus, if we define $\phi(\zeta) = \zeta^{r-1}\omega(\zeta)$ we have, using (12) and (13),

$$|J_n'(x)| \leqslant M_r C_{r-1} \int_0^\pi \phi\left(\zeta_n(t) + \frac{|u|}{n}\right) k_{n,r+1}(u)\, du. \tag{14}$$

The case $s = 0$ of Lorentz [**66**, Chapter 5, Lemma 1] is valid even though it is not claimed; the same method of proof is applicable. Applying this to (14) gives (10) for the case $\nu = 1$ since $t = \operatorname{arc}\cos x$ and $\Delta_n \leqslant 4\Delta_{n+1}$. For the remaining cases of (10) we need merely apply Theorem 3 of Lorentz [**66**, Chapter 5] to (10) repeatedly. This completes the proof of Theorem 12.2. □

The only rather specialized result which we will need about moduli of continuity is the following.

LEMMA 12.4. *Suppose $f \in C^r([a, b])$, $g \in C^{r+1}([a, b])$, and $f^{(r)}$ is not constant on $[a, b]$. Then*

$$\omega((fg)^{(r)}, h) \leqslant C\omega(f^{(r)}, h)$$

for some number C not depending on h.

PROOF. Since $f^{(r)}$ is not a constant

$$\lim_{h\to 0} \omega(f^{(r)}, h)/h > 0$$

(Timan [**60**, §3.2.6]); hence there exist $h_0 > 0$ and $C_0 > 0$ such that

$$\omega(f^{(r)}, h)/h \geqslant C_0, \qquad h \leqslant h_0. \tag{15}$$

Using Liebniz's rule we have

$$(fg)^{(r)} = \sum_{s=0}^{r} \binom{r}{s} f^{(r-s)} g^{(s)} = f^r g + F$$

where $F \in C^1([a, b])$. From the mean value theorem there are constants C_1 and C_2 such that $\omega(g, h) \leqslant C_1 h$, $\omega(F, h) \leqslant C_2 h$; hence ($\|\cdot\| = \|\cdot\|_{[a,b]}$)

$$\begin{aligned}
\omega((fg)^{(r)}, h) &\leqslant \omega(f^{(r)}g, h) + \omega(F, h)\\
&\leqslant \|g\|\omega(f^{(r)}, h) + \|f^{(r)}\|\omega(g, h) + \omega(F, h)\\
&\leqslant \|g\|\omega(f^{(r)}, h) + (\|f^{(r)}\|C_1 + C_2)h\\
&\leqslant (\|g\| + (\|f^{(r)}\|C_1 + C_2)/C_0)\omega(f^{(r)}, h), \qquad h \leqslant h_0,
\end{aligned}$$

where the last inequality uses (15). Also $\omega((fg)^{(r)}, \cdot)/\omega(f^{(r)}, \cdot)$ is continuous hence bounded on $[h_0, b-a]$ and the result follows. □

From Proposition 5.9 we know that there exists an integral polynomial $\tilde{q}$ with $\|\tilde{q}\|_{[a,b]} < 1$ and

$$J([a,b]) = Z_{\tilde{q}} \cap [a,b]. \tag{16}$$

Squaring $\tilde{q}$, if necessary, we have $0 \leqslant \tilde{q} < 1$ and it is obvious that $\deg \tilde{q} > 0$. Set $\tilde{\tilde{q}}(x) = \prod_{j=1}^{k}(x - x_j)$. We claim that to prove the sufficiency of the condition (2) of Theorem 12.1, we can assume without loss of generality that $\deg \tilde{\tilde{q}} \geqslant 1$. If $\deg \tilde{\tilde{q}} = 0$ then J is empty; hence $[a, b]$ contains no integers. Thus for some integer a' and $b' = a' + 1$ we have $a' < a < b < b'$. We extend f to $\tilde{f}$ on $[a', (b + b')/2]$ as follows. First extend $f^{(r)}$ to g on the interval $[(a' + a)/2, (b + b')/2]$ by specifying $q = f^{(r)}(a)$ on $[(a' + a)/2, a]$ and $g = f^{(r)}(b)$ on $[b, (b + b')/2]$. Clearly $\omega(g, \cdot) = \omega(f^{(r)}, \cdot)$ on $[0, b - a]$. Let $\tilde{f}$ be an rth indefinite integral of g with the constants of integration chosen so as to make $\tilde{f} \equiv f$ on $[a, b]$. Let φ be any infinitely differentiable function on $\mathbf{R}$ such that $\varphi \equiv 1$ on $[a, b]$ and $\varphi \equiv 0$ off of $[(a + a')/2, (b + b')/2]$. Then $\varphi\tilde{f}$ extends in the obvious way to $[a', (b + b')/2]$ so as to be identically zero near a. Furthermore, by Lemma 12.4, $\omega((\varphi f)^{(r)}, \cdot) \leqslant C\omega(\tilde{f}^{(r)}, \cdot) = C\omega(f^{(r)}, \cdot)$ and our claim is established.

LEMMA 12.5. *If p_n is a polynomial with degree $\leqslant n$ and divisible by $\tilde{q}^s$, s a positive integer, then there is an integral polynomial q_n with degree $\leqslant n$ and satisfying*

$$\|p_n - q_n\|_{[a,b]} = O(n^{-s}). \tag{17}$$

PROOF. By Lemma 4.4 we can represent p_n in the form

$$p_n = \sum_{j=s}^{N} A_j \tilde{q}^j \tag{18}$$

where the A_j's are polynomials satisfying $\deg A_j < \deg \tilde{q}$ $(s \leqslant j \leqslant N)$ and $N = [n/\deg \tilde{q}]$. It is easy to see that each power x^j $(s \leqslant j \leqslant N)$ can be written as a linear combination (with integral coefficients) of the polynomials $x^i(1 - x)^{N-i}$ $(s \leqslant i \leqslant N)$. Substituting $\tilde{q}$ for x therein and using (18) we have

$$p_n = \sum_{j=s}^{N} B_j \tilde{q}^j (1 - \tilde{q})^{N-j}$$

where $\deg B_j < \deg q$ $(s \leqslant j \leqslant N)$. Set

$$q_n = \sum_{j=s}^{N} [B_j] \tilde{q}^j (1 - \tilde{q})^{N-j}$$

where $[B_j]$ is the polynomial obtained from B_j by replacing each of its coefficients by its integral part. Because of the bound on the degrees of the B_j we have that $\|B_j - [B_j]\|_{[a,b]} \leqslant C$ for some constant C depending only on the interval $[a, b]$. Then on $[a, b]$:

$$
\begin{aligned}
|p_n - q_n| &= \left| \sum_{j=s}^{N} (B_j - [B_j])\tilde{q}^j (1 - \tilde{q})^{N-j} \right| \\
&\leqslant C \sum_{j=s}^{N} \tilde{q}^j (1 - \tilde{q})^{N-j} \\
&= C \sum_{j=s}^{N-s} \tilde{q}^j (1 - \tilde{q})^{N-j} + C \sum_{j=N-s+1}^{N} \tilde{q}^j (1 - \tilde{q})^j \\
&\leqslant C \binom{N}{s}^{-1} \sum_{j=s}^{N-s} \binom{N}{j} \tilde{q}^j (1 - \tilde{q})^{N-j} + C \sum_{j=N-s+1}^{N} \tilde{q}^j \\
&\leqslant C \binom{N}{s}^{-1} + sC \|\tilde{q}\|^{N-s+1};
\end{aligned}
$$

hence we have $\|p_n - q_n\| = O(1/N^s) = O(1/n^s)$. Since replacing B_j by $[B_j]$ does not raise the degree, we are done. □

Lemma 12.6. *For any three positive integers $s < m < n$ there exist a constant C not depending on n and real numbers a_j and b_j $(m \leqslant j \leqslant n)$ such that*

$$
\left\| x^s - \sum_{j=m}^{n} a_j x^j \right\|_{[-\lambda,\lambda]} \leqslant C\lambda^s / n^s \tag{19}
$$

and

$$
\left\| x^s - \sum_{j=m}^{n} b_j x^j \right\|_{[0,\lambda]} \leqslant C\lambda^s / n^{2s}. \tag{20}
$$

Furthermore, if s is even (resp., odd), then $a_j = 0$ for every odd (resp., even) index j.

Proof. We will prove (20) and indicate the changes necessary to get (19). By the obvious change of variable we need only consider the case $\lambda = 1$. From von Golitschek [70, Lemma 2] there exist real numbers b_j $(m \leqslant j \leqslant n)$ such that

$$
\left\| x^s - \sum_{j=m}^{n} b_j x^j \right\|_{[0,1]} \leqslant \prod_{j=m}^{n} \frac{j-s}{j+s} \leqslant \prod_{j=m}^{n} \exp(-2s/j) = \exp\left(-2s \sum_{j=m}^{n} j^{-1} \right)
$$

where the second inequality follows from the inequality $(1 - x)/(1 + x) \leqslant e^{-2x}$ $(x \geqslant 0)$ applied factorwise after dividing through by j and this, in turn, is proved by elementary methods. But

$$
\sum_{j=m}^{n} j^{-1} \geqslant \int_m^{n+1} dt/t = \ln((n+1)/m); \tag{21}
$$

hence $\exp(-2s\sum_{j=m}^{n} j^{-1}) \leqslant ((n+1)/m)^{-2s} = O(n^{-2s})$ which establishes (20). To obtain (19) first suppose that s is even. Then it suffices to approximate x^s on $[0, \lambda]$ using only the even exponents between m and n. The same lemma applies again only we must estimate $\sum_{j=m; j \text{ even}}^{n} j^{-1}$. This can be done as in (21) except that a factor of $\frac{1}{2}$ appears since only essentially half the summands are present, compared to before. The case where s is odd is similar. □

We will need the following strengthening of the result in Lemma 12.5. First factor $\tilde{q}$ (see (16)) as follows. If $x_0 \in J = J_0$ ($[a, b]$) then the minimal polynomial $m(x)$ of x_0 divides $\tilde{q}$. We can thus factor out the product $\tilde{q}_1$ of the minimal polynomials associated with J such that

$$\tilde{q} = \tilde{q}_1\tilde{q}_2 \tag{22}$$

and $\tilde{q}_2 \neq 0$ on $[a, b]$. Since each minimal polynomial is monic and integral ($\in \mathbf{Z}[x]$), so is $\tilde{q}_1$. It is clear then, from an inspection of the division algorithm, that $\tilde{q}_2$ is also integral.

LEMMA 12.7. *If p_n is a polynomial of degree $\leqslant n$ which is divisible by $\tilde{\tilde{q}}^s$, s a positive integer, then there exists an integral polynomial q_0 with degree at most $n + s \deg \tilde{q}_2$ and satisfying*

$$\|\tilde{q}_2^s p_n - q_0\|_{[a,b]} = O(n^{-s}).$$

PROOF. Since $\deg \tilde{\tilde{q}} \geqslant 1$ we can invoke Lemma 4.4 to write

$$p_n = \sum_{j=s}^{N} A_j \tilde{\tilde{q}}^j \tag{23}$$

where each A_j is a polynomial with degree $\leqslant \deg \tilde{\tilde{q}}$ ($s \leqslant j \leqslant N$) and $N = [n/k]$ ($k = \deg \tilde{\tilde{q}}$). Pick an integer l such that $\tilde{q}_1 | \tilde{\tilde{q}}^l$. Designate by $[A_j]$ the polynomial obtained from A_j by replacing each coefficient by its integral part ($s \leqslant j \leqslant sl$). Then for $s \leqslant j \leqslant sl$

$$\|[A_j] - A_j\|_{[a,b]} \leqslant C \tag{24}$$

where C depends only on the interval $[a, b]$. From Lemma 12.5 we see that for $s \leqslant j \leqslant sl - 1$ there are real numbers $a_{i,j}$, $ls \leqslant i \leqslant N - 1$, such that

$$\left| x^j - \sum_{i=ls}^{N-1} a_{i,j} x^i \right| = O((N-1)^{-j}) = O(N^{-j}), \qquad |x| \leqslant \|\tilde{\tilde{q}}\|_{[a,b]};$$

hence

$$\left\| \tilde{\tilde{q}}^j - \sum_{i=ls}^{N} a_{i,j} \tilde{\tilde{q}}^i \right\|_{[a,b]} = O(N^{-j}) = O(N^{-s}) = O(n^{-s})$$

since $N = [n/k]$. From (23) and (24)

$$p_n = \sum_{j=s}^{N} A_j \tilde{\tilde{q}}^j = \sum_{j=s}^{ls-1} [A_j] \tilde{\tilde{q}}^j + \sum_{j=ls}^{N} B_j \tilde{\tilde{q}}^j + O(n^{-s}).$$

Then

$$\tilde{q}_2^s p_n = \tilde{q}_2^s \sum_{j=s}^{ls-1} [A_j] \tilde{\tilde{q}}^j + \tilde{q}_2^s \sum_{j=ls}^{N} B_j \tilde{\tilde{q}}^j + O(n^{-s}). \tag{25}$$

We can apply Lemma 12.5 to the second summand on the right-hand side of (25) since it is divisible by $\tilde{q}^s$ and approximate it by an integral polynomial of degree $\leqslant n + s \deg \tilde{q}_2$ and with an error of the order $O((n + s \deg \tilde{q}_2)^{-s}) = O(n^{-s})$. □

LEMMA 12.8. *Let $\alpha \neq 0 \neq \beta$ and s, m, n be positive integers, $n > s + 2m$. There is a polynomial p of degree $\leqslant n_0 + n/2$ such that, for $x \in [-\lambda, \lambda]$,*

$$|x^s - x^{s+2m}p(x^2)| \leqslant C\frac{(|x^2 - \alpha^2|\,|x^2 - \beta^2|)^{[r/2]+1}}{n^s}$$

where $C = C(\lambda, m, r)$ does not depend on x or n and $n_0 = n_0(m, r, s)$.

PROOF. By equation (19) of Lemma 12.6 for $\tilde{C} = \tilde{C}(\lambda)$ we have $|x^s - x^{s+2m}p(x^2)| \leqslant \tilde{C}/n^s$ where p is a polynomial of degree $\leqslant (n - s - 2m)/2$. Multiplying both sides by

$$\left\{\left(\left(\frac{x}{\alpha}\right)^{2m} - 1\right)\left(\left(\frac{x}{\beta}\right)^{2m} - 1\right)\right\}^{[r/2]+1} = 1 + cx^{2m} + \cdots + dx^{4m([r/2]+1)}$$

$$= \left((x^2 - \alpha^2)(x^2 - \beta^2)\tilde{\tilde{p}}(x)\right)^{[r/2]+1}$$

gives

$$|x^s - x^{s+2m}\tilde{p}(x^2)| \leqslant \tilde{C}\frac{(|x^2 - \alpha^2|\,|x^2 - \beta^2|)^{[r/2]+1}}{n^s}(\tilde{\tilde{p}}(x))^{[r/2]+1}$$

$$\leqslant C\frac{(|x^2 - \alpha^2|\,|x^2 - \beta^2|)^{[r/2]+1}}{n^s}$$

where $C = \tilde{C}\|(\tilde{\tilde{p}}(x))^{[r/2]+1}\|_{[a,b]}$ and

$$\deg \tilde{p} \leqslant (n + 4m([r/2] + 1) - s - 2m)/2 \leqslant n_0 + n_2. \quad \square$$

We now turn to the proof of Theorem 12.1. We can assume without loss of generality that (2) can be replaced by the condition

$$f^{(\nu)}(x_j) = 0, \qquad 0 \leqslant \nu \leqslant r,\ 1 \leqslant j \leqslant k, \tag{26}$$

as follows. If $f^{(r)}$ is a constant, then f and q are identical and the result is obvious. If not, then as in the proof of Lemma 12.4 there exist $h_0 > 0$ and $C > 0$ such that $\omega(h)/h > C$ for $0 < h \leqslant h_0$ and then by continuity, $\omega(h)/h > C$ for all h. Thus

$$\begin{aligned}\omega((f - q)^{(r)}, h) &\leqslant \omega(f^{(r)}, h) + \omega(q^{(r)}, h)\\ &\leqslant \omega(f^{(r)}, h) + \|q^{(r+1)}\|_{[a,b]}h\\ &\leqslant \omega(f^{(r)}, h) + C^{-1}\|q^{(r+1)}\|_{[a,b]}\omega(f^{(r)}, h)\\ &= O(\omega(f^{(r)}, h)).\end{aligned}$$

Assuming the theorem to be true in case (26) holds, we apply it with $f - q$ in place of f and obtain q_n such that

$$\begin{aligned}|(f - q)^{(\nu)}(x) - q_n^{(\nu)}(x)| &\leqslant C_r\tilde{\Delta}_n^{r-\nu}(x)\omega((f - q)^{(r)}, \tilde{\Delta}_n(x))\\ &\leqslant C_r'\tilde{\Delta}_n^{r-\nu}(x)\omega(f^{(r)}, \tilde{\Delta}_n(x))\end{aligned}$$

and the general result follows for $n \geqslant (r+1)k$ by taking $(q_n + q)$ in place of q_n. It then follows for all n by modifying the constant C_r'. Thus we assume (26) from now on.

From Corollary 12.3 we have $(0 \leqslant \nu \leqslant r)$

$$\left|(f/\tilde{q}_2^{2r+2})^{(\nu)} - p_n^{(\nu)}(x)\right| \leqslant C_r \tilde{\Delta}_n^{r-\nu}(x)\tilde{\omega}(\tilde{\Delta}_n(x))$$

where $\deg p_n \leqslant n$ and $\tilde{\omega}(\cdot) = \omega((f/\tilde{q}_2^{2r+2})^{(r)}, \cdot)$. Since $\tilde{q}_2$ is a polynomial with no roots in $[a, b]$ we can apply Lemma 12.4 to obtain $(0 \leqslant \nu \leqslant r)$

$$\left|(f/\tilde{q}_2^{2r+2})^{(\nu)} - p_n^{(\nu)}(x)\right| \leqslant C_r \tilde{\Delta}_n^{r-\nu}(x)\omega(\tilde{\Delta}_n(x)) \tag{27}$$

with a change in the constant C_r.

From Liebniz's rule and (26) we see that

$$(f/\tilde{q}_2^{r+2})^{(\nu)}(x_j) = 0, \qquad 0 \leqslant \nu \leqslant r, 1 \leqslant j \leqslant k, \tag{28}$$

and by Corollary 12.3 we can assume that $p_n^{(\nu)}(x_j) = 0$, $0 \leqslant \nu \leqslant r$, $1 \leqslant j \leqslant k$. Equivalently, $\tilde{\tilde{q}}^{r+1}$ divides p_n and applying Lemma 4.4 to the quotient $p_n/\tilde{\tilde{q}}^{r+1}$ and then multiplying through by $\tilde{\tilde{q}}^{r+1}$ we obtain

$$p_n = \sum_{s=r+1}^{N_1} A_s \tilde{\tilde{q}}^s \tag{29}$$

where $\deg A_s < k$ $(r + 1 \leqslant s \leqslant N_1)$ and $N_1 = [n/k]$.

We now consider four cases according to the values of $\tilde{\tilde{q}}(a)$ and $\tilde{\tilde{q}}(b)$.

Case I. $\tilde{\tilde{q}}(a) \neq 0 \neq \tilde{\tilde{q}}(b)$. Let $r + 1 \leqslant s < 2r + 1$. In Lemma 12.8 we choose $\alpha = \tilde{\tilde{q}}(a)$, $\beta = \tilde{\tilde{q}}(b)$, $m \geqslant (2r + 2 - s)/2$ and substitute $\tilde{\tilde{q}}(x)$ for x to obtain

$$\begin{aligned}\left|\tilde{\tilde{q}}^s(x) - \tilde{\tilde{q}}^{s+2m}(x)p(\tilde{\tilde{q}}^2(x))\right| &\leqslant C\frac{\left(|\tilde{\tilde{q}}^2(x) - \tilde{\tilde{q}}^2(a)|\,|\tilde{\tilde{q}}^2(x) - \tilde{\tilde{q}}^2(b)|\right)^{[r/2]+1}}{n^s} \\ &\leqslant \tilde{C}\frac{((x-a)(b-x))^{[r/2]+1}}{n^{r+1}}\end{aligned} \tag{30}$$

where $\tilde{C} = C\|(\tilde{\tilde{q}}^2(x) - \tilde{\tilde{q}}^2(a))(\tilde{\tilde{q}}(x) - \tilde{\tilde{q}}^2(b))/(x-a)(x-b)\|_{[a,b]}$. From (29), using the same technique as in the proof of Lemma 12.7, we can write

$$p_n = \sum_{s=r+1}^{2r+1} [A_s]\tilde{\tilde{q}}^s + \sum_{s=r+1}^{2r+1} (A_s - [A_s])\tilde{\tilde{q}}^s + \sum_{s=2r+2}^{N_1} A_s\tilde{\tilde{q}}^s \tag{31}$$

where $\|A_s - [A_s]\|_{[a,b]} = O(1)$. Using (30) to substitute for $\tilde{\tilde{q}}^s$ in the second term of the right-hand side of (31) gives

$$\begin{aligned}p_n(x) = &\sum_{s=r+1}^{2r+1} [A_s](x)\tilde{\tilde{q}}^s(x) + \sum_{s=2r+2}^{N_2} A_s(x)\tilde{\tilde{q}}^s(x) \\ &+ O\left(\frac{((x-a)(b-x))^{[r/2]+1}}{n^{r+1}}\right)\end{aligned}$$

where $N_2 \leqslant n + 2n_0 + 2r + 1 + 2m = n + n_1$ and n_1 is independent of n.

Making use of (15), or rather its extension to all h, we can write

$$p_n = \sum_{s=r+1}^{2r+1} [A_s]\tilde{\tilde{q}}^s + \sum_{s=2r+2}^{N_2} A_s\tilde{\tilde{q}}^s + O\left(\left(\frac{\sqrt{(x-a)(x-b)}}{n}\right)^r \omega\left(\frac{\sqrt{(x-a)(b-x)}}{n}\right)\right). \tag{32}$$

We show how to establish this result in the other cases and then the theorem follows quickly.

Case II. $\tilde{\tilde{q}}(a) = \tilde{\tilde{q}}(b) = 0$. By hypothesis $x - a$ and $b - x$ both divide $\tilde{\tilde{q}}(x)$ and setting $\varphi(x) = \tilde{\tilde{q}}(x)/(x-a)(b-x)$ we have $\|\varphi\|_{[a,b]} < \infty$. Let $r + 1 \leqslant s \leqslant 2r + 1$ and $m \geqslant (2r + 2 - s)/2$. From Lemma 12.6 we have

$$|x^s - x^{s+2m}p(x^2)| \leqslant Cn^{-s} \quad \text{for } |x| \leqslant \|\varphi\|_{[a,b]}.$$

Substituting $\varphi(x)$ for x we have, on $[a, b]$,

$$|\varphi^s - \varphi^{s+2m}p(\varphi^2)|_{[a,b]} \leqslant Cn^{-s}.$$

Multiplying both sides by $((x-a)(b-x))^{s/2}$ we have

$$|\tilde{\tilde{q}}^s(x) - \tilde{\tilde{q}}^{s+2m}(x)p(\tilde{\tilde{q}}^2(x))| = O\left(\left(\frac{\sqrt{(x-a)(b-x)}}{n}\right)^s\right) = O\left(\left(\frac{\sqrt{(x-a)(b-x)}}{n}\right)^{r+1}\right).$$

Using this estimate in place of (30) we can proceed as in Case I to establish (32).

Case III. $\tilde{\tilde{q}}(a) \neq 0 = \tilde{\tilde{q}}(b)$. Here we may simply combine the techniques in Cases I and II. First use Lemma 12.8 with $\alpha = \beta = \tilde{\tilde{q}}(a)$, then substitute $(\tilde{\tilde{q}}(x)/\sqrt{b-x}\,)$ into the resulting inequality, and finally multiply both sides by $(b-x)^{s/2}$.

Case IV. $\tilde{\tilde{q}}(a) = 0 \neq \tilde{\tilde{q}}(b)$. We simply note that this is the same as Case III with the end points reversed.

Thus, in all cases we have (32), from which we can write

$$\tilde{q}_2^{2r+2}p_n = \tilde{q}_2^{2r+2} \sum_{s=r+1}^{2r+1} [A_s]\tilde{\tilde{q}}^s + \tilde{q}_2^{2r+2} \sum_{s=2r+2}^{N_2} A_s\tilde{\tilde{q}}^s + O\left(\left(\frac{\sqrt{(x-a)(b-x)}}{n}\right)^r \omega\left(\frac{\sqrt{(x-a)(b-x)}}{n}\right)\right). \tag{33}$$

By Lemma 12.7 there is an integral polynomial with degree $\leqslant n + n_1 + (2r + 2)\deg \tilde{q}_2 \leqslant n + n_2$ which approximates the middle term in the right-hand side of (33) to within $O(1/n^{2r+2})$. Adding this to the first term in (33) we obtain an integral polynomial q_N with degree $\leqslant n + n_2$ which satisfies

$$|\tilde{q}_2^{2r+2}(x)p_n(x) - q_N(x)| = O\left(\left(\frac{\sqrt{(x-a)(b-x)}}{n}\right)^r \omega\left(\frac{\sqrt{(x-a)(b-x)}}{n}\right) + \frac{1}{n^{2r+2}}\right). \quad (34)$$

Using (15) we have

$$\left(\frac{\sqrt{(x-a)(b-x)}}{n}\right)^r \omega\left(\frac{\sqrt{(x-a)(b-x)}}{n}\right) + \frac{1}{n^{2r}}\frac{1}{n^2}$$
$$\leqslant \left\{\max\left\{\frac{\sqrt{(x-a)(b-x)}}{n}, \frac{1}{n^2}\right\}\right\}^r \left(\omega\left(\frac{\sqrt{(x-a)(b-x)}}{n}\right) + \frac{1}{C_0}\omega\left(\frac{1}{n^2}\right)\right)$$
$$= O(\tilde{\Delta}_n^r(x)\omega(\tilde{\Delta}_n(x))).$$

Combining this with (34) we have

$$|\tilde{q}_2^{2r+2}(x)p_n(x) - q_N(x)| = O(\tilde{\Delta}_n^r(x)\omega(\tilde{\Delta}_n(x))) \quad (35)$$

where $\deg q_N \leqslant n + n_2$.

It remains to show that q_N is the desired polynomial. Set $p = \tilde{q}_2^{2r+2}$. From (35) and a result of Brudnyĭ (see Lorentz [**66**, Chapter 5, Theorem 3]), it follows that for $0 \leqslant s \leqslant r$

$$|(p(x)p_n(x))^{(s)} - q_N^{(s)}(x)| \leqslant \tilde{C}_r(\tilde{\Delta}_n^{r-s}(x)\omega(\tilde{\Delta}_n(x)))$$
$$= \tilde{C}_r\psi_{n,r-s}(x),$$

the last equality serving to define $\psi_{n,r-s}$. Using Liebniz's rule, the last inequality and (27) may be written in the forms

$$\left|\sum_{j=0}^{s}\binom{s}{j}p_n^{(j)}(x)p^{(s-j)}(x) - q_N^{(s)}(x)\right| \leqslant \tilde{C}_r\psi_{n,r-s}(x) \quad (36)$$

and

$$\left|\sum_{j=0}^{s}\binom{s}{j}f^{(j)}(x)(p^{-1}(x))^{(s-j)}(x) - p_n^{(s)}(x)\right| \leqslant C_r\psi_{n,r-s}(x), \quad (37)$$

respectively. Substituting the approximations of the $p_n^{(j)}$'s in (37) in place of the occurrences of the $p_n^{(j)}$'s in (36) $(0 \leqslant j \leqslant s)$ we obtain

$$\left|\sum_{j=0}^{s}\binom{s}{j}p^{(s-j)}(x)\sum_{l=0}^{j}\binom{j}{l}f^{(l)}(x)(p^{-1}(x))^{(j-l)} - q_N^{(s)}(x)\right| \leqslant \tilde{\tilde{C}}_r\psi_{n,r-s}(x).$$

Reversing the order of summation gives

$$\left|\sum_{l=0}^{s}f^{(l)}(x)\sum_{j=l}^{s}\binom{s}{j}\binom{j}{l}p^{(s-j)}(x)(p^{-1}(x))^{(j-l)} - q_N^{(s)}(x)\right| \leqslant \tilde{\tilde{C}}_r\psi_{n,r-s}(x).$$

But

$$\sum_{j=l}^{s}\binom{s}{j}\binom{j}{l}p^{(s-j)}(x)(p^{-1}(x))^{(j-l)} = \binom{s}{l}\sum_{j=l}^{s}\binom{s-l}{j-l}(p^{-1}(x))^{(j-l)}(p(x))^{(s-j)}$$

$$= \binom{s}{l}(p^{-1}(x)p(x))^{(s-l)} = \begin{cases} 1, & l = s, \\ 0, & l < s; \end{cases}$$

hence $|f^{(s)}(x) - q_N^{(s)}(x)| \leqslant \tilde{\tilde{C}}_r\tilde{\Delta}_n^{r-s}(x)\omega(\tilde{\Delta}_n(x))$. Since $\deg q_N \leqslant n + n_2$ where n_2 does not depend on n and

$$\tilde{\Delta}_{n-n_2}^{r-s}(x)\omega\big(\tilde{\Delta}_{n-n_2}(x)\big) = O\big(\tilde{\Delta}_n^{r-s}(x)\omega(\Delta_n(x))\big),$$

Theorem 12.1 is completely proved. □

EXAMPLE 12.9. In the case where $[a, b] = [0, 1]$ we can see as in Example 11.7 that for f to be approximable as in Theorem 12.1, it is necessary and sufficient that the quantities $\{f^{(j)}(0)/j!\}_{j=0}^{r}$ and $\{f^{(j)}(1)/j!\}_{j=0}^{r}$ be integers.

It is possible to sharpen the estimate in Theorem 12.1 in case $[a, b] = [-1, 1]$ and $\nu = 0$ as follows. We use the same notation as in that theorem.

THEOREM 12.10. *There are polynomials q_n in $\mathbf{Z}[x]$ with $\deg q_n \leqslant n$ and a C which depends only on f, r, a, and b such that, for $|x| \leqslant 1$ and $n > r$,*

$$|f(x) - q_n(x)| \leqslant C\left(\frac{\sqrt{1-x^2}}{n}\right)^r\omega\left(\frac{\sqrt{1-x^2}}{n}\right) \tag{38}$$

if and only if there is a q in $\mathbf{Z}[x]$ satisfying

$$q^{(\nu)}(x) = f^{(\nu)}(x), \qquad 0 \leqslant \nu \leqslant r, \quad x = 0, \pm 1. \tag{39}$$

PROOF. The necessity of condition (39) follows from Theorem 12.1 once we establish that $J[-1, 1] = \{-1, 0, 1\}$. The latter result is a consequence of the results in Example 3.15.

Conversely, suppose that condition (39) holds. We know from Theorem 12.1 that there are polynomials $\tilde{q}_n$ in $\mathbf{Z}[x]$ with $\deg \tilde{q}_n \leqslant n$ and such that for $|x| \leqslant 1$ and all n we have

$$|f(x) - \tilde{q}_n(x)| \leqslant C\left(\max\left\{\frac{\sqrt{1-x^2}}{n}, \frac{1}{n^2}\right\}\right)^r\omega\left(\max\left\{\frac{\sqrt{1-x^2}}{n}, \frac{1}{n^2}\right\}\right)$$

$$\leqslant C\left(\frac{\sqrt{1-x^2}}{n} + \frac{1}{n^2}\right)^r\omega\left(\frac{\sqrt{1-x^2}}{n} + \frac{1}{n^2}\right).$$

Let r_n be the sequence of Hilbert interpolating polynomials defined by the conditions

$$r_n^{(\nu)}(\pm 1) = f^{(\nu)}(\pm 1) - \tilde{q}_n^{(\nu)}(\pm 1), \qquad 0 \leqslant \nu \leqslant [r/2]. \tag{40}$$

We will show that the polynomials $q_n = r_n + \tilde{q}_n$ have the desired properties. We first note that from the theory of interpolating polynomials $\deg r_n \leqslant 2[r/2] + 1 \leqslant r + 1$; hence for $n > r$ we have $\deg q_n = \deg(r_n + \tilde{q}_n) \leqslant n$, as required.

We next establish that the q_n lie in $\mathbf{Z}[x]$. We need only show that r_n is in $\mathbf{Z}[x]$ since we already know that $\tilde{q}_n$ is in $\mathbf{Z}[x]$. From (39) and (40) we have

$$r_n^{(\nu)}(\pm 1) = q^{(\nu)}(\pm 1) - q_n^{(\nu)}(\pm 1), \qquad 0 \leqslant \nu \leqslant [r/2]. \tag{41}$$

By the division algorithm we can write

$$q(x) - \tilde{q}_n(x) = (x^2 - 1)^{([r/2]+1)} w(x) + \tilde{r}_n(x) \tag{42}$$

where $\deg \tilde{r}_n \leqslant 2[r/2] + 1$. By differentiating (42) (ν times) we see that $\tilde{r}_n^{(\nu)}(\pm 1) = q^{(\nu)}(\pm 1) - \tilde{q}_n^{(\nu)}(\pm 1)$, $0 \leqslant \nu \leqslant [r/2]$. From this and (41) we see that r_n and its derivatives up to and including order $[r/2]$ coincide with $\tilde{r}_n$ and its corresponding derivatives. In view of the bounds on the degrees, we see that r_n and $\tilde{r}_n$ are identical. It suffices then to show that $\tilde{r}_n \in \mathbf{Z}[x]$. This follows from the fact that it is obtained by dividing an integral polynomial $(q - \tilde{q}_n)$ by a monic integral polynomial $(x^2 - 1)$ as a cursory inspection of the division algorithm shows.

It remains to show that the estimate (38) is satisfied with the present choice of q_n. This has already been done in the proof of Theorem 2 in Teljakovskiĭ **[66]**. (His notation for the present $r_n(x)$ is $p(n, x)$.) □

PART IV: HISTORICAL NOTES AND REMARKS

HISTORICAL NOTES AND REMARKS

Introduction. The result in formula (6) appeared for the first time in Kantorovič [**31**]. The idea of representing the polynomial of best approximation with real coefficients in terms of the Bernšteĭn polynomials and then approximating the coefficients by integers is due to Bernšteĭn [**30**] and [**36**]. See also Kuz'min [**36**].

In Chlodovsky [**25**] the question was first posed as to whether or not the coefficients of the approximating polynomials could be taken from a more general set than **Z**. It is obvious that using the present method (approximating the coefficients of the Bernšteĭn polynomials) for proving the approximability of functions on [0, 1] which are zero at both 0 and 1 we can relax the condition on the set E of coefficients of our approximating polynomials to the condition that there exists a number $\delta > 0$ such that every interval of length δ contains at least one member of E. This was pointed out by Gagaeff [**29**] who based his argument on Chlodovsky's method, however. Other papers in this direction are Gilenko [**53**], Havinson [**59**], Gel'fond [**66**], Stafney [**67**], Samokis [**63**], Roulier ([**70**] and [**72**]), and Trigub [**77**].

The determination of the monic polynomials of least supremum norm on intervals was first carried out by Čebyšev [**1859**].

Chapter 2. The concept of Čebyšev polynomials was introduced in Čebyšev [**1859**] wherein they are calculated explicitly for intervals. Proposition 2.3 goes back to Haar [**18**] in the case of real valued functions and to Kolmogorov [**48**] in the complex case. The present proof is that of Rivlin and Shapiro [**60**]. The concept of transfinite diameter was introduced in Fekete [**23**] where he proves equality between the transfinite diameter and the Čebyšev constant.

A good general reference for this chapter is Hille [**62**]. See also Tsuji [**59**] for a more extensive treatment.

In connection with Theorem 2.11 we mention that the problem of finding the monic polynomial of degree n with least norm on an interval, say, has a rather long history. In the case of the L_1 norm, it was essentially solved in Stieltjes [**1876**]. Indeed, applying his formula (8) we obtain $M = S = 2^{-n}$; hence, for the interval $[-1, 1]$ the norm is $2 \cdot 2^{-n} = 2^{1-n}$ (the same as for the uniform norm).

The result can be obtained for other intervals by the usual change of variables. Also, from the fact that he knew the zeros (see his formula (6)), he could easily have calculated the polynomial explicitly. In the L_1, L_2 and uniform norms see Timan [**84**, §§2.9.31, 2.9.32 and 2.9.1 (respectively)] for the explicit determination of the polynomials with least norm on the interval $[-1, 1]$.

Chapter 3. The notion of the algebraic kernel was introduced by Fekete in [**54**], with respect to the ring of integers of an algebraic number field. With respect to an arbitrary discrete subring of **C** with rank 2 it appears in Ferguson [**65**] (see also Ferguson [**68a**]). It is explicitly calculated for the intervals $[a, b]$ for which the interval $[a + c, b + c]$ is contained in $[-2, 2]$ for some choice of an integer c, in Hewitt and Zuckerman [**20**]. For arbitrary subsets of the plane which can be translated by an integral amount into a subset of the unit circle the algebraic kernel is calculated in Ferguson [**70a**]. Earlier results, concerning whether or not the algebraic kernel is finite, are found in Fekete and Szegő [**55**], Schur [**18**] and Robinson [**62**].

Example 3.12 goes back to Fekete and Szegő [**55**, § 1.4].

The technique of proof in Lemma 3.6 is due to Okada [**23**, §3].

Chapter 4. This case was first considered by Fekete who assumed the coefficients to be the algebraic integers of an imaginary quadratic field. The results were announced in Fekete ([**54a**], [**54b**]) and detailed proofs given in Fekete [**55**] which is in Hebrew. When the integers are assumed to form an arbitrary discrete subring of **C** with rank 2, the results were first given in Ferguson [**65**]. (See also Ferguson [**68a**].)

The result in Proposition 4.2 first appeared in Laverent'ev [**36**]. It is an immediate consequence of Mergelyan's theorem. A modern proof of the latter can be found in Rudin [**74**]. See also the present Theorem 7.14 and remarks following it.

Theorem 4.9 is due to David Cantor [**69**].

For extensions of the results of this chapter to approximation by integral rational functions with poles among the rational integers (or at ∞), see D. Cantor [**75**].

In Trigub [**71**] the result in our Theorem 4.8 is claimed without any restriction on A other than that it be discrete and of rank 2. This is false as the following simple example shows. Let $L = Q(\sqrt{-1})$ so that I_L is the ring of Gaussian integers $\mathbf{Z} + i\mathbf{Z}$. It is easy to see that $A = Z + 2i\mathbf{Z}$ is a discrete subring of I_L with rank 2. Let $X = \{i\}$. Then the function $f(x) = x - (1 + 2i)$ is interpolated on $J_0(X) = X = \{i\}$ by itself, which is an element of $A[z]$. Obviously f is A-approximable on X but the unique Lagrange interpolating polynomial to f on $J_0(X)$ is the constant polynomial $(-1 - i)$ which is not in $A[z]$. The same error appears in Ferguson [**68a**]. (The false step in the proof is that the monic polynomial with simple roots at the points of $J_0(X, A)$ is not necessarily an element of $A[z]$; hence division by it can take one outside of $A[z]$.)

Concerning Lemma 4.3 the following comments may be of interest. The first result here seems to be in Fekete [**23**, Theorem XIV] who proves it for rational integral polynomials on sets X which are symmetric relative to the real axis. (Even earlier, for the L_2 norm, rational integral coefficients and on intervals of length less then 4 it was established by Hilbert [**1894**].) Fekete's method is capable of handling the present case as well. However, the proof we have given is more elementary. The method is due to Kakeya who used it on the intervals of the form $[-\alpha, \alpha]$, $\alpha < 2$ (see Okada [**23**, §3]). (See also Fukasawa [**26a**].) The extension of Fekete's method to the estimation of norms of minimal integral polynomials in several real variables on parallelopipeds in Euclidean spaces is given in Žirnova [**58**].

Besides the transfinite diameter, there is another very interesting constant which we can associate with a compact subset of **C** which reflects, in a sense, its "size". By the same argument as in Proposition 2.8 we can prove that

$$\lim_{n\to\infty} \|q_n^\circ\|_X^{1/n} = \tilde{d}(X)$$

exists, where each q_n° is an integral polynomial of degree $\leqslant n$ and minimal norm $\|\cdot\|_X$. From Fekete's method it follows that $\tilde{d} \leqslant \sqrt{d}$ (Trigub [**49**, §3.1]). Also, if a_n is the leading coefficient of q_n°, then $|a_n| \geqslant 1$; hence $p_n = a_n^{-1}q_n^\circ$ is monic and $\|p\|_X \leqslant \|q_n^\circ\|_X$. If $d(X) < 1$ then $q_n^\circ \to 0$; hence $\{\deg q_n^\circ\}$ is unbounded. Thus the numbers $\|p_n\|^{1/n}$ bound a certain subsequence of $\{\|t_n\|^{1/n}\}$ (the Čebyšev polynomials) and we conclude that $d \leqslant \tilde{d}$ $(d < 1)$. Unlike $d(X)$ the quantity $\tilde{d}(X)$ may change when X is translated.

It seems that I. G. Schnirelman and A. O. Gel′fond [**46**] first observed the following interesting connection between $\tilde{d}([0, 1])$ and the distribution of prime numbers. For each positive integer n let Ω_n denote the least common multiple of $\{1, 2, \ldots, n\}$ and let q_n° be a rational integral polynomial of degree $\leqslant n$ and deviate the least from zero among all rational integral polynomials of degree $\leqslant n$ $(\|q_n^\circ\| = \inf\|q_n\|)$ on $[0, 1]$. Then

$$\Omega_{2n+1}\|q_n\|^2 \geqslant \Omega_{2n+1}\int_0^1 q_n^2(x)\,dx \geqslant 1$$

since the middle quantity is obviously a positive integer. It is easy to see that $\pi(n)\ln n \geqslant \ln \Omega_n$ for all n. Thus, for every $\varepsilon > 0$,

$$\frac{\pi(2n+1)\ln(2n+1)}{2n} \geqslant \ln \frac{1}{\tilde{d}+\varepsilon}$$

for large n; hence

$$\lim_{n\to\infty} \frac{\pi(n)\ln(n)}{n} \geqslant \ln \frac{1}{\tilde{d}}$$

and the prime number theorem would follow quickly if we could show that $\tilde{d}^{-1} = e$. However, it is now known that $\tilde{d}^{-1} < e$. Interest remains in this approach in higher dimensions, however. See Trigub [**49**, Chapter 3] for more details.

For integral polynomials with minimum supremum norm, see also Sanov [**49**].

Chapter 5. The fundamental paper with the general results here is Hewitt and Zuckerman [**59**]. The fact that these results follow from those in the complex case seems to have been first recognized by D. Cantor (private conversation).

Both statements of Theorem 5.10 were established in the case of the L_2 metric by Aparicio [**55**].

Chapter 7. Theorem 7.21 is due to Alper [**64**]. The present proof is somewhat less complicated than his. See also Trigub [**71**]. The complex case for dimensions greater than 1 was first considered in Ferguson [**65**]. (See also Ferguson [**68b**].) For Theorem 7.24 see Ferguson [**69**].

The class of Mergelyan sets is wider than we have shown here. In Petrosjan [**70a, b**] we see that it includes the Weil polyhedra. More generally, it includes the strictly pseudoconvex regions (Kerzeman [**70**], [**71**], Lieb [**69**], Henkin [**69**]).

Chapter 8. The first half of the proof of Theorem 8.1 (sufficiency of the condition $\sum_{k=1}^{\infty}\lambda_k^{-1} = \infty$) is partially adapted from Korevaar [**75**]. The second half is found in Rudin [**74**].

Müntz' theorem was generalized by Szász [**16**] to cover the case of complex exponents with positive real parts. His condition for density is $\sum_{k=1}^{\infty}\mathrm{Re}\,\lambda_k/(1 + |\lambda_k|^2) = \infty$ and for nondensity $\sum_{k=1}^{\infty}(1 + \mathrm{Re}\,\lambda_k)/(1 + |\lambda_k|^2) < \infty$. See also the proof in Paley and Wiener [**34**]. Note that the two conditions are not exhaustive. The question of a necessary and sufficient condition in the case of complex exponents seems to be open. See Siegel [**72**].

Müntz' theorem is an extension of Weierstrass' theorem since the latter is obtained by setting $\lambda_k = k$, $1 \leqslant k$. Another extension of Weierstrass' theorem is that of Jackson where the degree of approximation of continuous functions by polynomials is studied. The analogue of this for integral polynomials is treated here in Chapter 12. A series of papers is devoted to the combination of these two extensions–so-called Müntz-Jackson theorems. The first is D. J. Newman [**65**] which treated the case of the L_2 norm. More recent papers are Bak and Newman [**72**], [**74**], Ganelius and Westlund [**70**], Leviatan [**74**], von Golitschek [**70**], [**73**], and others. Most of the results to date are special cases of those in von Golitschek [**76a**]. A survey of the results up to 1972 is available in Newman [**73**] and von Golitschek [**73**]. Finally, the case of a Müntz-Jackson theorem for polynomials with integral coefficients is treated in von Golitschek [**76b**].

Papers on the possibility of approximation by Λ-polynomials, in reverse chronological order, are Ferguson and von Golitschek [**75**] and [**74**], Martirosjan [**73**] and Ferguson [**74b**]. This is an area of current research activity. One can ask what happens to each of the known theorems in the classical case ($\Lambda = \mathbf{N}$). In Aparicio [**55**] an estimate is obtained for the minimum weighted L_2 norm on an interval $[0, \alpha]$ of Λ-polynomials with rational integral coefficients where the elements of Λ are real. See also Sanov [**29**] in the case of the uniform norm. Later, in Aparicio [**62**] the results are extended to the case of integral Λ-polynomials with complex exponents and coefficients from the integers of an imaginary quadratic field. In Aparicio [**55**] it is shown that if $\lambda_n^{-2}\sum_{k=1}^n \lambda_k \to \infty$ as $n \to \infty$ then for any $\alpha > 0$ and any $\varepsilon > 0$ there exists a rational integral Λ-polynomial q with $\int_0^\alpha q^2(x)\,dx < \varepsilon$. This is the case, then, if we take $\Lambda = \{n^\gamma\}_{n=1}^\infty$ with a fixed γ, $0 < \gamma < 1$. This is in contrast with the case $\Lambda = \mathbf{N}$ where we know from Theorem 2.11 that this is impossible for $\alpha \geqslant 4$.

For estimates of the error of best approximation for approximation by integral Λ-polynomials, see von Golitschek [**76**].

Chapter 9. The case of the L_2 norm and Lebesgue measure on an interval of length < 4 in our Theorem 9.11 forms the main result in Aparicio [**55**].

The case of the approximation of a function of several real variables by polynomials with integral coefficients was first treated in Fukasawa [**26b**]. See also Håstad [**58**], Žirnova [**58**] and Hewitt and Zuckerman [**59**]. Another approach appears in Yamamoto [**31**] but the argument is inconclusive.

Chapter 10. Existence and uniqueness questions for approximation by integral polynomials were first considered in Andria [**68**]. There the set X is any interval of length less than 4 and the integral polynomials are the elements of $\mathbf{Z}[x]$. (See also Andria [**71**].)

We do not have existence when approximating by the ring of all integral polynomials (without a bound on the degree) since, as we see from our general results, this ring is not closed in general and there obviously cannot exist a best approximation to a function which is in the closure but not in the ring itself. Andria observes this in [**71**], where he also shows that a function outside the closure of the ring does not necessarily have a best approximation from this ring.

In a later paper, Ferguson [**74a**], these results were extended to the general, real and complex cases. Also the existence question, when approximating by polynomials of finite degree, is considered there for the first time.

The present Theorem 10.9 is an extension of Andria [**73**, Theorem 1] who proves it where X is an interval of $\mathbf{R}$ of length less than 4 and $Z = \mathbf{Z}$. The present proof is simpler as well as more general than his. Andria goes on to apply this result to obtain one on approximation in L_p norms.

One further remark on Andria [**73**]. His Theorem 3 is a consequence of the general results in Ferguson [**68a**] since the set X there is not assumed to be infinite and all norms on function spaces on finite sets are equivalent. Furthermore, it follows from this that Andria's set C_ω can be replaced by the (generally) smaller set $J(W)$, thus strengthening his Theorem 3.

There are obvious connections between the subject of this book and diophantine approximation. See Trigub [**71**], Kovalevskaja [**73**] and the bibliography of the latter.

Chapter 11. The first consideration of this problem seems to be in Bernšteĭn [**30**]. Theorem 11.5 solves a problem posed in Bernšteĭn [**36**]. It first appears in Trigub [**61**] and [**62**]. See also Trigub [**71**]. Theorem 11.2, its proof and Theorem 11.3 appear in Trigub [**71**]. The present proof of Theorem 11.3 is shorter than the original. A special case ($q(z) \equiv z$) is the fact that a power series with integral coefficients and radius of convergence greater than one is a polynomial (Pólya [**23**]). The first results of the type of Theorem 11.4 are to be found in Gel'fond [**55**] for the intervals [0, 1] and [−1, 1] where an estimate is given for ρ. Sufficiency in the general case here appeared in Trigub [**71**]. For the explicit arithmetic conditions in Example 11.7 when [0, 1] is replaced by [−1, 1] see Gel'fond [**55**, Theorem IV].

Chapter 12. For asymptotic results in the case of L_p norms, see Trigub [**71**]. Theorem 12.2 first appeared in Gel'fond [**55**] for the interval [0, 1]. In its present generality it appears in Trigub [**61**] and [**62**]. The first paper with a quantitative result of this type is Okada [**24**] whose proof seems to be in error, however. The first correct result of this type is that in Kantorovič [**31**] which we have reproduced in the introduction.

For quantitative results in the approximation of functions of several real variables, see Žirnova [**58**], Trigub [**71**], and Kovalevskaja [**73**].

In the case of arbitrary coefficients the estimate in Theorem 12.1 is essentially due to Timan [**51**], the difference being that the present $\varepsilon_n = \max\{\sqrt{(x-a)(b-x)}\,/n, 1/n^2\}$ is replaced by $(\sqrt{(x-a)(b-x)}\,/n) + (1/n^2)$. Much later, Timan's result for the case of arbitrary coefficients was established with ε_n replaced by simply $\sqrt{(x-a)(b-x)}\,/n$. This seems to be independently due to Gopengauz [**67**], Runck [**68**], and Teljakovskiĭ [**66**]. The present Theorem 12.10 is new and answers the question of a similar improvement in the case of integral coefficients and approximation on [−1, 1]. Another interesting open problem is that of removing the dependency of the constant C on the approximated function f. In the case of approximation by polynomials with arbitrary coefficients, C is independent of f.

APPENDIX

APPROXIMATION AT ALGEBRAIC INTEGERS

The purpose of this appendix is to establish some fundamental theorems on approximation by integral polynomials on finite sets of algebraic integers. The first three derive from a conversation with David Cantor. These are essential in the more general situation. Throughout, the symbol z will be used only to represent an element of $\mathbf{C}$ and n-tuples of complex numbers will be written out in full. Also, A will be any discrete subring of $\mathbf{C}$ of rank 2 and L the unique imaginary quadratic field such that $A \subset I_L$ (Proposition 1.10).

The first theorem tells us that we can approximate by integral polynomials on any incomplete set of conjugate algebraic integers over the imaginary quadratic field containing A. In the case of a complete set of conjugate algebraic integers, the situation is just the opposite; we can approximate only what we can interpolate. This is a corollary of Proposition 3.7.

THEOREM A.1. *Let $\alpha_1, \ldots, \alpha_n$ be a complete set of conjugate algebraic integers over L, ε any positive number, and $z_2, \ldots, z_n$ any complex numbers. Then there is a polynomial $q \in A[z]$ such that*

$$|q(\alpha_j) - z_j| < \varepsilon, \qquad 2 \leqslant j \leqslant n.$$

PROOF. From Proposition 1.11 we know that there is a positive rational integer m such that $mI_L \subset A$. If the theorem were true for I_L in place of A we could find $q_0 \in I_L[z]$ satisfying

$$|q_0(\alpha_j) - z_j/m| < \varepsilon/m, \qquad 2 \leqslant j \leqslant m$$

from which the conclusion of the general theorem follows if we take $q = mq_0$. Thus we can assume from the outset that $A = I_L$.

By Proposition 1.7, I_L is a discrete subring of $\mathbf{C}$ of rank 2. Then $(I_L)^n$ is a discrete subgroup of $\mathbf{C}^n$ with rank $2n$. We identify $\mathbf{C}^n$ with $\mathbf{R}^{2n}$ by the map $(z_1, \ldots, z_n) \to (\operatorname{Re} z_1, \operatorname{Im} z_1, \ldots, \operatorname{Re} z_n, \operatorname{Im} z_n)$. Thus, by Cassels [**59**, Theorem VI, p. 78], $(I_L)^n$ is a lattice in $\mathbf{R}^{2n}$. Consider the complex linear transformation T of $\mathbf{C}^n$ represented by the matrix

$$\begin{bmatrix} 1 & \alpha_1 & \cdots & \alpha_1^{n-1} \\ 1 & \alpha_2 & \cdots & \alpha_2^{n-1} \\ & & \cdots & \\ 1 & \alpha_n & \cdots & \alpha_n^{n-1} \end{bmatrix} \tag{1}$$

with respect to the canonical basis of $\mathbf{C}^n$ as a complex linear space. It is clear from definitions that T is also a real linear transformation of the $2n$-dimensional real linear space underlying $\mathbf{C}^n$. We denote this real linear transformation by the different symbol T_0, since, in general, we will have $\det T \neq \det T_0$. Set

$$c_1 = 2^{n-1/2}(\det T_0)\delta((I_L)^n)$$

and $c_2 = c_3 = \cdots = c_{2n} = 2^{-1/2}$, where $\delta((I_L)^n)$ denotes the determinant of the lattice $(I_L)^n$ (Cassels [**59**, p. 10]). Then by Cassels [**59**, Theorem III, p. 73] there exists $(y_1, \ldots, y_n)$ in $(I_L)^n$ such that $(y_1, \ldots, y_n) \neq (0, \ldots, 0)$ and

$$|\mathrm{Re}(w_j)| < 2^{-1/2}, \quad |\mathrm{Im}(w_j)| < 2^{-1/2}, \qquad 2 \leqslant j \leqslant n,$$

where $(w_1, \ldots, w_n) = T(y_1, \ldots, y_n)$. Consequently,

$$|w_j|^2 = |\mathrm{Re}(w_j)|^2 + |\mathrm{Im}(w_j)|^2 < 1, \qquad 2 \leqslant j \leqslant n.$$

If we define $\tilde{q}(z) = y_1 + y_2 z + \cdots + y_n z^{n-1}$ then we have, by the representation (1) for T,

$$|\tilde{q}(\alpha_j)| = |w_j| < 1, \qquad 2 \leqslant j \leqslant n. \tag{2}$$

Since L has characteristic zero, the minimal polynomial over L of the set of conjugates $\alpha_1, \ldots, \alpha_n$ is separable, i.e., the α's are distinct. Since the determinant of (1) is $\Pi_{1 \leqslant i < j \leqslant n}(\alpha_j - \alpha_i)$, T is a nonsingular transformation. Thus, since $(y_1, \ldots, y_n) \neq (0, \ldots, 0)$, we have $w_j = \tilde{q}(\alpha_j) \neq 0$ for some j. Since the α's are conjugate to α_j, this implies that

$$\tilde{q}(\alpha_j) \neq 0, \qquad 1 \leqslant j \leqslant n.$$

Now suppose that p is any polynomial with complex coefficients and degree $< n - 1$. If $[p]$ denotes a polynomial obtained by replacing each coefficient of p by a nearest integer (i.e., element of I_L) we have

$$|([p] - p)(\alpha_j)| \leqslant \sum_{m=0}^{n-2} \delta|\alpha_j^m|, \qquad 2 \leqslant j \leqslant n, \tag{3}$$

where δ is defined in Proposition 1.2. By (2) there is a positive integer k such that

$$\left(\sum_{m=0}^{n-2} \delta|\alpha_j^m|\right)|\tilde{q}(\alpha_j)|^k < \varepsilon, \qquad 2 \leqslant j \leqslant n. \tag{4}$$

Let p be the Lagrange interpolating polynomial such that $p(\alpha_j) = z_j/\tilde{q}^k(\alpha_j)$, $2 \leqslant j \leqslant n$. Then $\deg p < n - 1$ so (3) and (4) give

$$\begin{aligned}|[p](\alpha_j)\tilde{q}^k(\alpha_j) - z_j| &= |[p](\alpha_j)\tilde{q}^k(\alpha_j) - p(\alpha_j)\tilde{q}^k(\alpha_j)| \\ &\leqslant |([p] - p)(\alpha_j)|\, |\tilde{q}(\alpha_j)|^k \\ &< \varepsilon, \qquad 2 \leqslant j \leqslant n.\end{aligned}$$

The conclusion of the theorem holds if we set $q = [p]\tilde{q}^k$. □

Utilizing some ideas from algebraic number theory, it is possible to give a shorter proof of Theorem A.1 as follows. We assume as before that $A = I_L$. Let p be the minimal polynomial of $\alpha_1, \ldots, \alpha_n$. Adjoin a root θ of p to L to get a field $F = L(\theta)$. Then p is the minimal polynomial of θ and p factors into linear factors over the completion $\hat{L} = \mathbf{C}$ of L with respect to the usual Archimedean valuation $|\cdot|$ on L. Let $|\cdot|_1, \ldots, |\cdot|_n$ be the extensions to F of $|\cdot|$ corresponding to the linear factors $x - \alpha_1, \ldots, x - \alpha_n$, respectively (Bachman [**64**, p. 133]). Then any element of $L(\theta)$ can be expressed in the form $p'(\theta)$ with $p' \in L[x]$ and $|p'(\theta)|_i = |p'(\alpha_i)|$.

There exists $b \in I_L$ such that $\theta_1 = b\theta$ is integral over I_L. Since $1, \theta, \ldots, \theta^{n-1}$ is a base for $L(\theta)$ over L, we see that $1, \theta_1, \ldots, \theta_1^{n-1}$ is also a base and consists of elements integral over I_L. Thus by Lang [**70**, Proposition 6, proof, p. 6] there exists $c \in I_L$, $c \neq 0$, such that $I_F \subset c^{-1}(I_L + I_L\theta_1 + \cdots + I_L\theta_1^{n-1})$; hence

$$I_F \subset c^{-1}(I_L + I_L\theta + \cdots + I_L\theta^{n-1}). \qquad (*)$$

Since L is dense in $\mathbf{C}$ we can assume that $z_2, \ldots, z_n \in L$. By the very strong approximation theorem (O'Meara [**63**, p. 77]) there exists $a \in F$ such that

$$|a - c^{-1}z_i|_i < |c|_i^{-1}\varepsilon, \qquad 2 \leqslant i \leqslant n,$$

and $|a|_\mathfrak{p} \leqslant 1$, $\mathfrak{p}$ non-Archimedean. Since $|a|_\mathfrak{p} \leqslant 1$ for all non-Archimedean $\mathfrak{p}$, $a \in I_F$. Thus, by $(*)$ there is a polynomial $q \in I_L[x]$ such that $c^{-1}q(\theta) = a$. Then

$$|c^{-1}q(\theta) - c^{-1}z_i|_i < |c|_i^{-1}\varepsilon, \qquad 2 \leqslant i \leqslant n;$$

hence $|q(\theta) - z_i|_i < \varepsilon$, $2 \leqslant i \leqslant n$, and $|q(\alpha_i) - z_i| < \varepsilon$, $2 \leqslant i \leqslant n$.

Our next result generalizes this theorem to the case of an arbitrary finite number of incomplete sets of conjugate algebraic integers.

THEOREM A.2. *Let*

$$\begin{matrix} \alpha_{11} & ,\ldots, & \alpha_{1r_1} \\ \alpha_{21} & ,\ldots, & \alpha_{2r_2} \\ & \cdots & \\ \alpha_{s1} & ,\ldots, & \alpha_{sr_s} \end{matrix}$$

be an array (not necessarily rectangular) with each row an incomplete set of conjugates integral over I_L and where the minimal polynomials $p_1, \ldots, p_s$ satisfied by the respective rows are distinct. If the array

$$\begin{matrix} z_{11} & ,\ldots, & z_{1r_1} \\ z_{21} & ,\ldots, & z_{2r_2} \\ & \cdots & \\ z_{s1} & ,\ldots, & z_{sr_s} \end{matrix}$$

consists of any complex numbers and $\varepsilon > 0$, *then there exists a* q *in* $A[z]$ *such that*

$$|q(\alpha_{ij}) - z_{ij}| < \varepsilon, \qquad 1 \leqslant i \leqslant s, 1 \leqslant j \leqslant r_i.$$

PROOF. Let $q_i' = \prod_{k \neq i} p_k$, $1 \leqslant i \leqslant s$. Then $q_i'(\alpha_{kl}) = 0$ if and only if $k \neq i$, by definition of the p's. For each i $(1 \leqslant i \leqslant s)$ there exists by Theorem A.1 a q_i'' in $A[z]$ such that

$$\left| q_i''(\alpha_{ij}) - \frac{z_{ij}}{q_i'(\alpha_{ij})} \right| < \frac{\varepsilon}{|q_i'(\alpha_{ij})|}, \qquad 1 \leqslant j \leqslant r_i.$$

Thus $|(q_i'' q_i')(\alpha_{ij}) - z_{ij}| < \varepsilon$, $1 \leqslant j \leqslant r_i$. If we set $q = q_1'' q_1' + \cdots + q_s'' q_s'$ then $q(\alpha_{ij}) = (q_i'' q_i')(\alpha_{ij})$ since $q_k'(\alpha_{ij}) = 0$ if $k \neq i$. Thus

$$|q(\alpha_{ij}) - z_{ij}| = |(q_i'' q_i')(\alpha_{ij}) - z_{ij}| < \varepsilon$$

for all i, j. □

We note in passing that another way of looking at Theorem A.2 is the following.

COROLLARY A.3. *If* $\{\alpha_1, \ldots, \alpha_k\}$ *is any set of algebraic integers which does not contain a complete set of conjugates over* L, *then the set of* k*-tuples* $\{(q(\alpha_1), \ldots, q(\alpha_k)) \colon q \in A[z]\}$ *is dense in* $\mathbf{C}^k$.

Another way in which Theorem A.1 may be generalized is to require that the derivatives at the given points also approximate. We needed this extension of the theorem in Chapter 7.

THEOREM A.4. *Let* $z_1, \ldots, z_n$ *be a complete set of conjugate (over* L*) algebraic integers in* $\mathbf{C}$ *where* L *is any imaginary quadratic field. Let* f *and* h *be defined and holomorphic functions in an open set containing* $\{z_2, \ldots, z_n\}$ *and*

$$|f(z_j)/h(z_j)| < \infty$$

for $2 \leqslant j \leqslant n$. *If* m *is any nonnegative integer and* $\varepsilon > 0$, *then there exists an integral polynomial* q *in* $A[z]$ *such that*

$$|f^{(\nu)}(z_j) - (hq)^{(\nu)}(z_j)| < \varepsilon, \qquad 0 \leqslant \nu \leqslant m, \quad 2 \leqslant j \leqslant n.$$

PROOF. We can assume without loss of generality that $A = I_L$ as follows. By Proposition 1.11 there is a positive integer m_0 such that $m_0 I_L \subset A$. If we knew the result for $A = I_L$ then we could find a $q \in I_L[z]$ satisfying

$$|f^{(\nu)}(z_j)/m_0 - (hq)^{(\nu)}(z_j)| < \varepsilon/m_0, \qquad 0 \leqslant \nu \leqslant m, \quad 2 \leqslant j \leqslant n;$$

hence

$$|f^{(\nu)}(z_j) - (hm_0q)^{(\nu)}(z_j)| < \varepsilon, \qquad 0 \leqslant \nu \leqslant m, \quad 2 \leqslant j \leqslant n,$$

and $m_0 q \in m_0 I_L[z] \subset A[z]$.

By Theorem A.1 there is a $\tilde{q} \in I_L[z]$ with $0 < |\tilde{q}(z_j)| < 1$, $2 \leqslant j \leqslant n$. Choose $\rho > 0$ so small that the disks of radius ρ and centers $\{z_2, \ldots, z_n\}$ are disjoint, lie within the domain of definition of f and h, and $0 < |q| < 1$ and $h \neq 0$ on $H = \bigcup_{j=2}^n (z_j + \rho D)$ where D is the closed unit disk. Let $\Delta = \max\{\nu! \rho^{-\nu}\}_{\nu=0}^m$.

Choose ε' so that $\Delta 2\varepsilon' < \varepsilon$ and then a positive integer λ such that

$$M\frac{\|h\|_H\|q\|_H^\lambda}{1 - \|q\|_H} < \varepsilon'$$

where M is defined by $M = (\deg q)\max_{0 \leqslant s < \deg q}\|x^s\|_H$. Then by a fundamental result due to Mergelyan (our Theorem 7.14) there is a polynomial p in $\mathbf{C}[z]$ such that

$$\|f/hq^\lambda - p\|_H < \varepsilon'/\|hq^\lambda\|_H;$$

hence

$$\|f - hq^\lambda p\|_H < \varepsilon'. \tag{$*$}$$

Utilizing Lemma 4.4 we can write $hq^\lambda p = h\Sigma_{k \geqslant \lambda} h_k q^k$ where the h_k's are polynomials satisfying $\deg h_k < \deg q$. If we define q by $q = \Sigma_{k \geqslant \lambda}[h_k]q^k$ where $[h_k]$ stands for h_k with each coefficient replaced by a nearest element of I_L then $q \in I_L[z]$ and

$$\|hq^\lambda p - hq\|_H \leqslant \|h\|_H M\|q\|_H^\lambda/(1 - \|q\|_H) < \varepsilon'.$$

From this and ($*$) we have $\|f - hq\|_H < 2\varepsilon'$. From the Cauchy integral formula we have for $0 \leqslant \nu \leqslant m$ and $2 \leqslant j \leqslant n$

$$\begin{aligned}|f^{(\nu)}(z_j) - (hq)^{(\nu)}(z_j)| &= \left|\frac{\nu!}{2\pi i}\int_{C_j}\frac{f(\xi) - (hq)(\xi)}{(\xi - z_j)^{\nu+1}}\,d\xi\right| \\ &\leqslant \nu!\rho^{-\nu}\|f - hq\|_H < \Delta 2\varepsilon' < \varepsilon\end{aligned}$$

where C_j is the circle with center z_j and radius ρ. □

THEOREM A.5. *The preceding theorem* (A.4) *holds whenever* $z_2, \ldots, z_n$ *is a set of algebraic integers which contains no complete set of conjugate algebraic integers over* L.

PROOF. We can assume without loss of generality that $A = I_L$ by the same argument as before. Let $S_1, \ldots, S_k$ be the decomposition of $z_2, \ldots, z_n$ under the equivalence relation of conjugacy over L and $q_1, \ldots, q_k$ the corresponding minimal polynomials. For each l $(1 \leqslant l \leqslant k)$ apply Theorem A.4 to get $\tilde{q}_l$ in $A[z]$ satisfying

$$\left|f^{(\nu)}(z_j) - \left(h\left(\prod_{j \neq l} q_j^m\right)\tilde{q}_l\right)^{(\nu)}(z_j)\right| < \varepsilon, \qquad 0 \leqslant \nu \leqslant m, \quad z_j \in S_l.$$

It suffices to take $q = \Sigma_{l=1}^k(\tilde{q}_l\Pi_{j \neq l}q_j^m)$ since for z_j in S_l and $0 \leqslant \nu \leqslant m$

$$(hq)^{(\nu)}(z_j) = \left(h\tilde{q}_l\prod_{j \neq l} q_j^m\right)^{(\nu)}(z_j). \quad □$$

BIBLIOGRAPHY

Al'per, S. Ya

[64] *On the approximation of functions by polynomials with integral coefficients on closed sets*, Izv. Akad. Nauk SSSR Ser. Math. **28** (1964), 1173–1186. (Russian) MR **30** #257.

Andria, D. George

[68] *On integral polynomial approximation*, Dissertation, St. Louis Univ., 1968.

[71] *Approximation of continuous functions by polynomials with integral coefficients*, J. Approximation Theory **4** (1971), 357–362. MR **44** #3055.

[73] *Convergence theorems for integral polynomial approximations*, J. Approximation Theory **7** (1973), 319–324. MR **50** #13979.

Aparicio Bernardo, Emiliano

[55] *On some properties of polynomials with integral coefficients and on the approximation of functions in the mean by polynomials with integral coefficients*, Izv. Akad. Nauk SSSR Ser. Mat. **19** (1955), 303–318. (Russian) MR **17**, 256.

[62] *On the least deviation from zero of quasipolynomials with algebraic integer coefficients*, Vestnik Moskov. Univ. Ser. I Mat. Meh. **1962**, no. 2, 21–32. (Russian) MR **25** #3917.

[67] *On the approximation of functions by polynomials with integer coefficients*, Proc. 8th Annual Reunion of Spanish Math. (Santiago, 1967) (Spanish), 21–33. Publ. Inst. 'Jorge Juan' Mat., Madrid, 1969. MR **41** #4066.

[72] *Metodo de las formas lineales para la acotacion de las desviaciones minimas a cero de los polinomios generalizados de coeficientes enteros*, Proc. 1st Conference of Portuguese and Spanish Mathematicians (Lisbon, 1972) 133–143. Inst. 'Jorge Juan' Mat., Madrid, 1973. MR **51** #3752.

[76] *Generalization of a theorem of M. Fekete to polynomials with integer coefficients in several unknowns*, Rev. Mat. Hisp.-Amer. (4) **36** (1976), 105–124. (Spanish).

Bachman, George

[64] *Introduction to p-adic numbers and valuation theory*, Academic Press, New York, 1964. MR **30** #90.

Bak, J. and Newman, D. J.

[72] *Müntz-Jackson theorems in $L^p[0, 1]$ and $C[0, 1]$*, Amer. J. Math. **94** (1972), 437–457. MR **46** #9605.

[74] *Müntz-Jackson theorems in L^p, $p < 2$*, J. Approximation Theory **10** (1974), 218–226. MR **50** #2755.

Bernšteĭn, S. N.

[30] *Some remarks on polynomials of least deviation with integral coefficients*, Dokl. Akad. Nauk SSSR (1930), 411–415 (Russian). Also: Sobranie Sočinenii: Tom I, Konstruktivnaya Teoriya Funktsii (1905–1930), Moscow, 1952, pp. 468–471, 562–563 (Russian). Translation: Collected Works: Vol. I, Constructive Theory of Functions (1905–1930), translated by the U. S. Atomic Energy Commission, AEC-tr-3460, pp. 140–144, 200–201.

[36] *The present state and some problems of the theory of approximation of functions of a real variable* *by polynomials*, Proc. First All-Union Math. Conf. (Khar'kov, 1930), M.-L., 1936, pp. 78–96 (Russian). Translation: Collected Works: Constructive Theory of Functions (1905–1930), translated by the U. S. Atomic Energy Commission, AEC-tr-3460, pp. 145–165.

Bourbaki, Nicolas

[63] *Éléments de mathématique. Fasc.* V. *Première partie*: *Les structures fondamentales de l'analyse. Livre* III. *Topologie générale*. Chapitres V–VIII, Actualités Sci. Indust., No. 1235, Hermann, Paris, 1963.

[64] *Éléments de mathématique. Fasc.* XXX. *Deuxième partie. Algèbre Commutative*. Chapitres V et VI, Actualitiés Sci. Indust., No. 1308, Hermann, Paris, 1964. MR **33** #2660.

Cantor, David G.

[65] *On the elementary theory of Diophantine approximation over the ring of adeles*. I, Illinois J. Math. **9** (1965), 644–700. MR **32** #5592.

[67] *On the Stone-Weierstrass approximation theorem for valued fields*, Pacific J. Math. **21** (1967), 473–478. MR **35** #1578.

[69] *On approximation by polynomials with algebraic integer coefficients*, Proc. Sympos. Pure Math., vol. XII, Amer. Math. Soc., Providence, R. I., 1969, pp. 1–13. MR **41** #1680.

[75] *On certain algebraic integers and approximation by rational functions with integral coefficients*, Pacific J. Math. **67** (1976), 323–338. MR **55** #2785.

Cassels, J. W. S.

[59] *An introduction to the geometry of numbers*, Die Grundlehren der Math. Wissenschaften, Band 99, Springer-Verlag, Berlin-Göttingen-Heidelberg, 1959. MR **28** #1175.

Čebyšev, P. L.

[59] *Sur les questions de minima qui se rattachent à la représentation approximative des fonctions*, Mém. de l'Académie Impériale des Sciences de St.-Pétersbourg, Sixième série **7** (1859), 199–291. Oeuvres, St. Petersburg, Vol. 1, pp. 271–378; reprint, Chelsea Publ. Co., New York, 1962.

Cheney, E. Ward

[66] *Introduction to approximation theory*, Internat. Series in Pure and Appl. Math., McGraw-Hill, New York, 1966. MR **36** #5568.

Chlodovsky, M. I.

[25] *Une remarque sur la représentation des fonctions continues par des polynomes à coefficients entiers*, Mat. Sb. **32** (1925), 472–474.

Fekete, Michael

[23] *Über die Verteiling der Wurzeln bei gewissen algebraischen Gleichungen mit ganzzahligen Koeffizienten*, Math. Z. **17** (1923), 228–249.

[54a] *Approximations par polynomes avec conditions diophantiennes*, C. R. Acad. Sci. Paris Sér. A.-B **239** (1954), 1337–1339. MR **16**, 694.

[54b] *Approximation par des polynomes avec conditions diophantinennes*. II, C. R. Acad. Sci. Paris Sér. A-B **239** (1954), 1455–1457. MR **16**, 694.

[55] *Approximation by polynomials with Diophantine side-conditions*, Riveon Lematematika **9** (1955), 1–12. (Hebrew, English summary) MR **17**, 477.

Fekete, M. and Szegö, G.

[55] *On algebraic equations with integral coefficients whose roots belong to a given point set*, Math. Z. **63** (1955), 158–172. MR **17**, 355.

Ferguson, Le Baron O.

[65] *Uniform approximations by polynomials with coefficients in discrete subrings of C*, Dissertation, Univ. of Washington, Seattle, 1965.

[68a] *Uniform approximation by polynomials with integral coefficients*. I, Pacific J. Math. **27** (1968), 53–69. MR **38** #4861.

[68b] *Uniform approximation by polynomials with integral coefficients*. II, Pacific J. Math. **26** (1968), 273–281. MR **38** #4861.

[69] *Uniform approximation of rational functions by polynomials with integral coefficients*, Duke Math. J. **36** (1969), 673–675. MR **40** #4462.

[70a] *Algebraic kernels of planar sets*, Duke Math. J. **37** (1970), 225–230. MR **41** #1673.

[70b] *Some remarks on approximation by polynomials with integral coefficients*, Approximation Theory (Proc. Sympos., Lancaster, July 1969), pp. 59–62, Academic Press, London and New York, 1970. MR **41** #8885.

[74a] *Existence and uniqueness in approximation by integral polynomials*, J. Approximation Theory **10** (1974), 237–244. MR **50** #10637.

[74b] *Müntz-Szász theorem with integral coefficients*. I, Functional Analysis and Its Applications Internat. Conf., Madras, 1973, Lecture Notes in Math., vol. 399, Springer-Verlag, New York, 1974, pp. 119–122. MR **55** #3623.

[76] *Approximation by integral Müntz polynomials*, Proc. Colloq. Fourier Analysis and Approximation Theory, Budapest, 1976, in Fourier Analysis and Approximation Theory, G. Alexits and P. Turán eds., North-Holland, Amsterdam, 1978.

[78] *Approximation by polynomials with integral coefficients and digital filter design*, Vortragsauszüge der Tagung über numerische Methoden der Approximationstheorie vom 18.–23. März 1979 im Mathematischen Forschungsinstitut Oberwolfach (Schwarzwald), Birkhäuser Verlag, Basel und Stuttgart, to appear.

Ferguson, Le Baron O. and Golitschek, M. von

[74] *Le théorème de Müntz-Szász avec coefficients entiers*, C. R. Acad. Sci. Paris **279** (1974), 817–818.

[75] *Müntz-Szász theorem with integral coefficients*. II, Trans. Amer. Math. Soc. **213** (1975), 115–126. MR **55** #3624.

Fukasawa, Seigo

[26a] *Über den Feketeschen Satz*, Tôhoku Math. J. **26** (1926), 201–204.

[26b] *Über die Näherungspolynome mit ganzzahligen Koeffizienten*, Tôhoku Math. J. **27** (1926), 267–270.

Gagaeff, M. B.

[29] *Sur la représentation des fonctions par des polynomes à coefficients qui appartiennent à un ensemble donné dénombrable*, Mat. Sb. **36** (1929), 184–187.

Ganelius, Tord and Westlund, S.

[70] *The degree of approximation in Müntz's theorem*, Topics in analysis (Colloq. Math. Anal. Jyväskylä 1970), pp. 125–132. Lecture Notes in Math., vol. 419, Springer, Berlin, 1974. MR **51** #8688.

Gel'fond, A. O.

[44] *Commentary on the two papers of P. L. Čebyšev "On the determination of the prime numbers not exceeding a given value" and "On prime numbers,"* The complete collected works of P. L. Čebyšev, Vol. 1, Theory of Numbers, Akad. Nauk SSSR, Moscow-Leningrad, 1944, pp. 285–288. (Russian) MR **6**, 254.

[55] *On uniform approximation by polynomials with rational integral coefficients*, Uspehi Mat. Nauk (N.S.) **10** (1955), no. 1 (63), 41–65. (Russian) MR **17**, 30. *Selected works*, Izdat. "Nauka", Moscow, 1973, p. 287.

[66] *On the approximation of polynomials with specially chosen coefficients*, Uspehi Mat. Nauk **21** (1966), no. 3 (129), 225–229. (Russian) MR **33** #6217. *Selected works*, Izdat. "Nauka", Moscow, 1973, p. 407.

Gilenko, N. D.

[53] *Representation of functions by series of polynomials with special coefficients*, Moskov. Gos. Ped. Inst. Učen. Zap. **71** (1953), 63–69. (Russian) MR **17**, 728.

Golitschek, Manfred von

[70] *Erweiterung der Approximationssätze von Jackson im Sinne von C. Müntz*, J. Approximation Theory **3** (1970), 72–76. MR **41** #2273.

[73] *Jackson-Müntz Sätze in der L_p-norm*, J. Approximation Theory **7** (1973), 87–106. MR **49** #5641.

[73] *Additional remarks to the paper of D. J. Newman*, Approximation Theory, G. G. Lorentz, ed., Proc. Internat. Sympos. (Austin, Texas, January, 1973), Academic Press, New York, pp. 213–217. MR **49** #3378.

[76a] *Jackson-Müntz-Szász theorems in $L^p[0, 1]$ and $C[0, 1]$ for complex exponents*, J. Approximation Theory **18** (1976), 13–29.

[76b] *Approximation durch Polynome mit ganzzahligen Koeffizienten*, Lecture Notes in Math., vol. 556, Springer-Verlag, New York, 1976, pp. 201–212.

[76c] *The degree of approximation for generalized polynomials with integral coefficients*, Trans. Amer. Math. Soc. **224** (1976), 417–425.

Gončarov, V. L.

[54] *Theory of interpolation and approximation of functions*, 2nd ed., Gosudarstv. Izdat. Tehn.-Teor. Lit. Moscow, 1954. (Russian) MR **16**, 803.

Gopengauz, I. E.

[67] *On a theorem of A. F. Timan on the approximation of functions by polynomials on a finite interval*, Mat. Zametki **1** (1967), 163–172. (Russian) MR **34** #8042. Transl.: Math. Notes **1** (1967).

Haar, A.

[1876] *Die Minkowskische Geometrie und die Annäherung and stetige Funktionen*, Math. Ann. **78** (1918), 294–311.

Hasse, Helmut

[63] *Zahlentheorie*, 2.auflage, Akademie-Verlag, Berlin, 1963. MR **27** #3621.

[64] *Vorlesungen über Zahlentheorie*, 2.auflage, Die Grundlehren der Math. Wissenschaften, Band 59, Springer-Verlag, Berlin, 1964. MR **32** #5569.

Håstad, Matts

[58] *Uniform approximation with Diophantine side-conditions of continuous functions*, Ark. Mat. **3** (1958), 487–493. MR **22** #12236.

Havinson, S. Ya.

[69] *Permissible magnitudes of the coefficients of polynomials in the uniform approximation of continuous functions*, Mat. Zametki **6** (1969), 619–625. (Russian) MR **41** #4068. Transl. Math. Notes **6** (1969), 834–838.

Henkin, G. M.

[69] *Integral representation of functions which are holomorphic in strictly pseudoconvex regions and some applications*, Mat. Sb. (N.S.) **78 (120)** (1969), 611–632. (Russian) Transl. Math. USSR-Sb. **7** (1969), 597–616. MR **40** #2902.

Hewitt, Edwin and Ross, K. A.

[63] *Abstract harmonic analysis*, Vol. 1, *Structure of topological groups. Integration theory, group representations*, Die Grundlehren der Math. Wissenschaften, Band 115, Academic Press, New York; Springer-Verlag, Berlin-Göttingen-Heidelberg, 1963. MR **28** #158.

Hewitt, Edwin and Zuckerman, H. S.

[59] *Approximation by polynomials with integral coefficients, a reformulation of the Stone-Weierstrass theorem*, Duke Math. J. **26** (1959), 305–324. MR **26** #6656.

Hilbert, D.

[1894] *Ein Beitrag zur Theorie des Legendre'schen Polynoms*, Acta Math. **18** (1894), 155–159.

Hille, E.

[62] *Analytic function theory, Vol.* II, *Introduction to higher mathematics*, Ginn and Co., New York, 1962. MR **34** #1490.

Hoffman, Kenneth

[62] *Banach spaces of analytic functions*, Prentice-Hall Series in Modern Analysis, Prentice-Hall, Englewood Cliffs, N. J., 1962. MR **24** #A2844.

Hörmander, L.

[66] *An introduction to complex analysis in several variables*, Univ. Series in Higher Math., D. Van Nostrand, Princeton, N. J., 1966. MR **34** #2933.

Jacobson, Nathan

[51] *Lectures in abstract algebra, Vol.* 1, *Basic concepts*, Univ. Series in Higher Math., D. Van Nostrand, Princeton, N. J., 1951. MR **12**, 794.

Kakeya, Sôichi

[14] *On approximate polynomials*, Tôhoku Math. J. **61** (1914), 182–186.

Kallin, Eva

[65] *Polynomial convexity*: *The three spheres problem*, Proc. Conf. Complex Analysis (Minneapolis, 1964), Springer-Verlag, New York, 1965, pp. 301–304. MR **31** #3631.

Kantorovič, L. V.

[31] *Some remarks on the approximation of functions by means of polynomials with integral coefficients*, Izv. Akad. Nauk SSSR Ser. Mat. (1931), 1163–1168. (Russian)

Kerzman, Norberto

[70] *Hölder and L^p estimates for solutions of $\bar\partial u = f$ in strongly pseudoconvex domains*, Bull. Amer. Math. Soc. **76** (1970), 860–864. MR **42** #4768.

[71] *Hölder and L^p estimates for solutions of $\bar\partial u = f$ in strongly pseudoconvex domains*, Comm. Pure Appl. Math. **24** (1971), 301–379. MR **43** #7658.

Kolmogorov, A. N.

[48] *A remark on the polynomials of P. L. Čebyšev deviating the least from a given function*, Uspehi Mat. Nauk **3** (1948), 216–221. (Russian) MR **10**, 35.

Korevaar, Jacob

[75] *Discrete sets of uniqueness for bounded holomorphic functions $f(z, w)$* (preprint).

Kovalevskaja, E. I.

[73] *Metric theorems on the approximation of zero by a linear combination of polynomials with integral coefficients*, Acta Arith. **25** (1973), 93–104. MR **49** #2633.

Kronecker, L.

[1857] *Zwei Sätze über Gleichungen mit ganzzahligen Koefficienten*, J. Reine Angew. Math. **53** (1857), 173–175.

Kuz′min, R. O.

[36] Proc. First All-Union Math. Conference (Khar′kov, 1930), M.-L., 1936.

Lang, Serge

[65] *Algebra*, Addison-Wesley Series in Mathematics, Addison-Wesley, London, 1965. MR **44** #5416.

[70] *Algebraic number theory*, Addison-Wesley, Reading, Mass., 1970. MR **44** #181.

Lavrent′ev, M. A.

[36] *Sur les fonctions d'une variable complex représentables par des séries de polynômes*, Actualités Sci. Indust., No. 441, Hermann, Paris, 1936.

Leviatan, D.

[74] *On the Jackson-Müntz theorem*, J. Approximation Theory **10** (1974), 1–5.

Lieb, Ingo

[69] *Ein Approximationssatz auf streng pseudokonvexen Gebieten*, Math. Ann. **184** (1969), 56–60. MR **41** #7146.

Lorentz, G. G.

[66] *Approximation of functions*, Holt, Rinehart and Winston, New York, 1966. MR **35** #4642; erratum, **36**, p. 1567.

Martirosian, V. A.

[73] *On uniform approximation by polynomials with respect to a Müntz system with integral coefficients*, Izv. Akad. Nauk Armjan. SSR Ser. Mat. **8** (1973), 167–175. (Russian. Armenian and English summaries) MR **48** #11852.

[75] *The possibility of uniform approximation by polynomials in the Müntz system with integral coefficients*, Izv. Akad. Nauk Armjan. SSR **10** (1975), 293–306, 386. (Russian. Armenian and English summaries) MR **53** #8724.

Mergelyan, S. N.

[51] *On the representation of functions by series of polynomials on closed sets*, Dokl. Akad. Nauk SSSR **78** (1951), 405–408. (Russian) MR **13**, 23. Amer. Math. Soc. Transl., no. 85, pp. 1–8.

Müller, Helmut

[75] *Bemerkungen zuer Approximation stetiger Funktionen in einer p-adischen Variablen*. I, J. Reine Angew. Math. **276** (1975), 167–169. MR **51** #10250.

[76] *Zur Approximation durch Polynome mit ganzen Koeffizienten*, J. Approximation Theory **16** (1976), 43–47. MR **53** #8725.

Naimark, M. A.

[64] *Normed rings*, revised ed., translated from the first Russian edition by Leo F. Boron, P. Noordhoff N. V., Groningen, 1964. MR **34** #4928; **22** #1824.

Newman, D. J.

[65] *A Müntz-Jackson theorem*, Amer. J. Math. **87** (1965), 940–944. MR **32** #4429.

[73] *A review of Müntz-Jackson theorems*, Approximation Theory, G. G. Lorentz, ed., Proc. Internat. Sympos. (Austin, Texas, January, 1973), Academic Press, New York, pp. 199–212. MR **49** #7671.

Okada, Yoshimoto

[23] *On approximate polynomials with integral coefficients only*, Tôhoku Math. J. **23** (1923), 26–35.

[24] *On the accuracy of approximation by polynomials with integral coefficients only*, Japan J. Math. **1** (1924), 29–31.

O'Meara, O. T.

[63] *Introduction to quadratic forms*, Die Grundlehren der Math. Wissenschaften, Band 117, Academic Press, New York; Springer-Verlag, Berlin-Göttingen-Heidelberg, 1963. MR **27** #2485.

Pál, Julius

[14] *Zwei kleine Bemerkungen*, Tôhoku Math. J. **6** (1914/1915), 42–43.

Paley, R. E. A. C. and Wiener, N.

[34] *Fourier transforms in the complex domain*, Amer. Math. Soc. Colloq. Publ., no. XIX, Amer. Math. Soc., Providence, R. I., 1934.

Petrosjan, A. I.

[70a] *Approximation by holomorphic functions on Weil polyhedra in the space* C^2, Izv. Akad. Nauk Armjan. SSR Ser. Mat. **5** (1970), 507–521. (Russian. Armenian and English summaries) MR **44** #6997.

[70b] *Uniform approximation of functions by polynomials on Weil polyhedra*, Izv. Akad. Nauk SSSR Ser. Mat. **34** (1970), 1241–1261. Transl.: Math. USSR-Izv. **4** (1970), 1250–1271. MR **44** #478.

Pólya, G.

[19] *Über ganzwertige Polynome in algebraischen Zahlköpern*, J. Reine Angew. Math. **149** (1919), 97–116.

[23] *Sur les séries entières à coefficients entiers*, Proc. London Math. Soc. **21** (1923), 22–38.

Rademacher, H.

[64] *Lectures on elementary number theory*, Blaisdell Publishing Co. [Ginn and Co.], New York, 1964. MR **30** #1079.

Rivlin, T. J. and Shapiro, H. S.

[60] *Some uniqueness problems in approximation theory*, Comm. Pure Appl. Math. **13** (1960), 35–47. MR **23** #A458.

Robinson, Raphael M.

[62] *Intervals containing infinitely many sets of conjugate algebraic integers*, Studies in Mathematical Analysis and Related Topics, Essays in Honor of George Pólya, Stanford Univ. Press, Stanford, Calif., 1962, pp. 305–315. MR **26** #2433.

Roulier, John A.

[70] *Permissible bounds on the coefficients of approximating polynomials*, J. Approximation Theory **3** (1970), 117–122. MR **41** #7345.

[72] *Restrictions on the coefficients of approximating polynomials*, J. Approximation Theory **6** (1972), 276–282. MR **49** #11101.

Rudin, Walter

[74] *Real and complex analysis*, 2nd ed., McGraw-Hill Series in Higher Math., McGraw-Hill, New York, 1974. MR **35** #1420.

Runck, P. O.

[68] *Bemerkungen zu den Approximationssätzen von Jackson und Jackson-Timan*, Abstract Spaces and Approximation (Proc. Conf. Oberwolfach, 1968), pp. 303–308. Birkhäuser, Basel, 1969. MR **42** #740.

Samokiš, B. A.

[63] *On the coefficients of polynomials approximating a continuous function uniformly on a segment*, Metody Vyčisl. **1** (1963), 58–65. (Russian) MR **32** #4431.

Sanov, I. N.

[49] *Functions with integral parameters, deviating the least from zero*, Leningrad. Gos. Univ. Učen. Zap. Ser. Mat. Nauk **111** (1949), 32–46.

Schur, I.

[18] *Über die Verteilung der Wurzeln bei gewissen algebraischen Gleichungen mit ganzzahligen Koeffizienten*, Math. Z. **1** (1918), 377–402.

Siegel, Alan R.

[72] *On the Müntz-Szász theorem for* $C[0, 1]$, Proc. Amer. Math. Soc. **36** (1972), 161–166. MR **46** #5899.

Stafney, James D.

[67] *A permissible restriction on the coefficients in uniform polynomial approximation to* $C[0, 1]$, Duke Math. J. **34** (1967), 393–396. MR **35** #3329.

Stark, H. M.

[67] *A complete determination of the complex quadratic fields of class number one*, Michigan Math. J. **14** (1967), 1–27. MR **36** #5102.

Stieltjes, T. J.

[1876] *Iets over de benanderde voorstelling van eene functie door eene andere*, Delft, 1876 (Dutch). French translation: *De la représentation approximative d'une fonction par une autre*, Oeuvres Complètes, Noordhoff, Groningen, Netherlands, 1914, pp. 11–20.

Szász, O.

[16] *Über die Approximation stetiger Funktionen durch lineare Aggregate von Potenzen*, Math. Ann. **77** (1916), 482–496.

Teljakovskiĭ, S. A.

[66] *Two theorems on approximation of functions by algebraic polynomials*, Mat. Sb. (N.S.) **70 (112)** (1966), 252–265. (Russian) MR **33** #1622; Amer. Math. Soc. Transl., Ser. 2, vol. 77, 1968.

Timan, A. F.

[51] *A strengthening of Jackson's theorem on the best approximation of continuous functions by polynomials on a finite segment of the real axis*, Dokl. Akad. Nauk SSSR **78** (1951), 17–20. (Russian) MR **12**, 823.

[60] *Theory of approximation of functions of a real variable*, Gosudarstv. Izdat. Fiz.-Mat. Lit., Moscow, 1960. (Russian) MR **22** #8257. Transl: Pure and Applied Math., vol. 34, Pergamon Press, New York, 1963. MR **33** #465.

Trigub, R. M.

[60] *Approximation of a function with given modulus of smoothness on the exterior of a segment and semiaxis*, Dokl. Akad. Nauk SSSR **132** (1960), 303–306. (Russian) MR **23** #A460.

[61] *Approximation of functions by polynomials with integral coefficients*, Dokl. Akad. Nauk SSSR **140** (1961), 773–775. (Russian) MR **24** #A1551.

[62] *Approximation of functions by polynomials with integral coefficients*, Izv. Akad. Nauk SSSR Ser. Mat. **26** (1962), 261–280. (Russian) MR **25** #373.

[63] *Approximation of functions by polynomials with integral coefficients*, Dissertation, Sumy, USSR, 1961. (Russian) Authors abstract, Sumy, USSR, 1963.

[71] *Approximation of functions with Diophantine conditions by polynomials with integral coefficients*, Metric Questions of the Theory of Functions and Mappings, No. 2, pp. 267–333. (Russian) Izdat. "Naukova Dumka", Kiev, 1971. MR **47** #683; **45** #509.

[77] *Approximation of functions by polynomials with special coefficients*, Izv. Vysš. Učebn. Zaved. Matematika **176** (1977), 93–99. (Russian) MR **56** #3307. Transl: Soviet Math. (Iz. VUZ) **21** (1977), 77–82.

Tsuji, M.

[59] *Potential theory in modern function theory*, Maruzen, Tokyo, 1959. MR **22** #5712.

Tzimbalario, J.

[77] *Approximation by generalized polynomials with integral coefficients*, preprint.

Walsh, J. L.

[63] *Interpolation and approximation by rational functions in the complex domain*, 3rd ed., Colloq. Publ., no. XX, Amer. Math. Soc., Providence, R. I., 1960. MR **36** #1672a.

Weiss, Edwin

[63] *Algebraic number theory*, Internat. Series in Pure and Applied Math., McGraw-Hill, New York, 1963. MR **28** #3021.

Wermer, J.

[61] *Banach algebras and analytic functions*, Advances in Math. **1** (1961), 51–102. MR **26** #629.

Yamamoto, I.

[31] *A remark on approximate polynomials*, Tôhoku Math. J. **33** (1931), 21–22.

Žirnova, G. A.

[58] *On the approximation of functions by polynomials with integral coefficients*, Izv. Vysš. Učebn. Zaved. Matematika **4** (1958), 80–88. (Russian) MR **24** #A949.

Rensselaer Polytechnic Institute

Université de Nancy

University of California, Riverside

University of California, San Diego

Stefan Banach International Mathematical Center, Warsaw

ABCDEFGHIJ–CM–89876543210